Life After Earth

The science of survival and the search for new worlds

Ellis Silver, PhD

ideas4writers

Published in Great Britain in 2026

by

ideas4writers
19 Crow Green
Cullompton
Devon
EX15 1EW

ideas4writers.com

ISBN 978-1-8383690-4-0

Cover design and layout by Dave Haslett

Life After Earth

Other books by Ellis Silver

Humans Are Not From Earth

SuperMars

Contents

Part VI – Legacy

Epilogue

Introduction: The clock is ticking

Earth will not last forever. Its habitability is finite – whether threatened by climate collapse, asteroid impact, nuclear war, runaway technology, or the slow but certain death of the Sun. For most of our history, such limits were abstract, far beyond the horizon of human concern. But now, they are visible.

This is not a book about despair. It is a road-map to what comes next. It explores where we might go when our home can no longer sustain us, how we might get there, and what we might become along the way. It is about the science that will carry us outward, the cooperation that will make it possible, the ethical challenges we will face, and the philosophical questions that will follow us to the stars.

Our journey begins with necessity: the urgent need to leave. It moves through the building of new worlds, the challenges of governing a multi-planetary species, and the possibilities of traveling beyond our Solar System. It imagines futures where we are no longer bound by biology, where we cross galaxies, and where we even survive the end of the universe. And it ends with the question of legacy – what remains of us when we are gone, and whether being remembered is the only measure of having mattered.

Part I

The necessity of departure

Dawn rises on a changing Earth: fires on the horizon, cities adapting to rising seas, satellites tracing cold arcs across the sky. We are a species that reads the weather and then builds ships. The choice before us is not whether to love our world; it is whether to love it enough to carry its story forward.

1

Why we must leave

Earth's habitability is limited by forces both natural and self-inflicted. Climate change, ecological collapse, asteroid impacts, supervolcanoes, pandemics, and the eventual expansion of the Sun into a red giant all threaten our long-term survival. Remaining bound to a single planet is a gamble we cannot win indefinitely. Some threats may be centuries away; others could arrive suddenly and without any warning.

Earth's future and the science of uninhabitability

For billions of years, Earth has been a cradle for life – a world where ecosystems have evolved, adapted, and endured through shifting climates and planetary transformations. But no planet remains habitable forever, and ours is no exception. Over time, natural cycles, solar changes, and cosmic forces will steadily erode Earth's ability to support life, driving an inevitable environmental decline.

Some of these changes will unfold over immense geological timescales. Others – including climate shifts we've triggered ourselves – could accelerate the process.

The uncomfortable truth is that we may face an uninhabitable Earth far sooner than most people imagine. We don't have billions of years to prepare. We don't even have *one* billion. Current estimates suggest that the Earth could cross the threshold into uninhabitability in less than 600 million years – and our work to prepare for it must begin now.

By studying the mechanisms that shape habitability, we can gain an insight into the challenges ahead and the possible solutions. These solutions do exist – but they are far beyond our present technological reach.

Preparing for these transformations – whether through environmental resilience, technological innovation, or expansion into space – is not just an intellectual exercise; it is a necessity for our species' survival.

1. The Sun's evolution

The Sun is not a static presence in our sky. It is a star in constant transformation – a "living" Sun – with a finite lifespan. Over the next few billion years, it will undergo changes that will reshape its structure and seal the fate of the planets that orbit it. Those shifts will happen slowly, but their consequences will be profound.

Increasing luminosity

As the Sun ages, hydrogen in its core fuses into helium, altering its internal balance. This process will steadily increase its energy output, making it brighter and more radiant.

Over the next 110 million years, the Sun's luminosity is expected to rise by about one percent. That might sound small, but the effects will cascade.

250 million years from now, the warming will become relentless, triggering climate changes severe enough to cause mass extinctions.

By 600 million years, carbon dioxide levels will have fallen so low that plants will no longer be able to survive, and ecosystems will collapse.

By 800 million years, oxygen levels could fall so drastically that over 99 percent of life will vanish.

Four billion years from now, the Sun's heat will be so intense that metals on Earth's surface could melt.

The decline of carbon dioxide

The Sun's growing brightness will speed up ocean evaporation, intensifying rainfall and accelerating the weathering of silicate rocks. This process – which is part of the carbonate-silicate cycle – draws carbon dioxide from the atmosphere and locks it into the crust.

As carbon dioxide levels fall, photosynthesis will falter. Within 600 to 800 million years, large-scale plant life will disappear, food chains will unravel, and mass extinctions will follow.

Oxygen loss and the collapse of complex life

Without plants, oxygen production will grind to a halt, and the biosphere will suffocate. Earth will become increasingly hostile to complex organisms. Eventually, the atmospheric balance that has sustained life for billions of years will collapse entirely, and the planet will become barren.

Ocean evaporation

As heat builds, the oceans will begin to vanish. Water vapor, which is a powerful greenhouse gas, will flood the atmosphere, trapping even more heat. This runaway greenhouse effect will accelerate until the oceans boil away completely – just as they did on Venus, where surface temperatures now exceed 842°F (450°C).

The Sun's red giant phase

In around 5.4 billion years from now, the Sun will exhaust its hydrogen fuel, marking the end of its stable main-sequence life and triggering dramatic changes. Without the outward pressure from fusion to counter gravity, the core will collapse and heat up, igniting helium fusion.

The intense heat from the contracting core will cause the outer layers to balloon outward, transforming the Sun into a red giant. Its atmosphere will then extend beyond the orbits of Mercury and Venus and – around 7.6 billion years from now – it may engulf Earth too.

Even if our planet escapes direct consumption, it will be stripped of its atmosphere and oceans, and reduced to a scorched, airless rock. This will occur long before the Sun's outer layers reach our neighborhood.

Toward the end of this phase, the Sun's core will become unstable, shedding its outer layers to form a planetary nebula. The remaining core will collapse into a dense white dwarf, which will slowly cool over trillions of years until it fades into darkness.

A universe in motion

The Sun's story is a reminder that nothing in the cosmos is permanent. Its transformations may span billions of years, but each one of them carries major consequences for Earth's habitability. Understanding these processes not only illuminates our planet's future but also deepens our knowledge of how stars live and die across the universe.

When we look at the Sun today, we are witnessing just one fleeting chapter in its vast, unfolding history – a history that will, in time, reshape and ultimately end the world we call home.

2. Magnetic field decay

Earth's magnetic field is one of the planet's most essential defenses. It acts as an invisible shield that protects us from harmful cosmic radiation and the constant stream of charged particles known as the solar wind. This protective barrier is generated deep underground by the motion of molten iron in the outer core – a process known as the geodynamo.

But this shield is not eternal. Over billions of years, it will gradually weaken, leaving Earth increasingly exposed to radiation and environmental decline.

The magnetic field naturally fluctuates, and even reverses polarity from time to time. But these events unfold over thousands to millions of years, giving life ample opportunity to adapt. The real danger lies much farther in the future.

In two to three billion years, the outer core is expected to cool and thicken, slowing the movement of liquid iron. As the geodynamo fades, the magnetic field will weaken and eventually vanish, stripping away one of Earth's most vital layers of protection. Even a partial loss could have profound consequences for the atmosphere, climate, and life itself.

Increased exposure to solar wind

The solar wind is a relentless stream of charged particles from the Sun. It is normally deflected by Earth's magnetic field, but without that shield, these particles would slam directly into the atmosphere, dramatically increasing radiation on the surface.

This exposure could disrupt Earth's atmospheric chemistry, accelerate climate change, and pose serious risks to living organisms, causing higher DNA mutation rates and radiation-related illnesses.

Atmospheric loss

Without this magnetic protection, the solar wind will slowly strip away Earth's atmosphere, carrying away essential gases such as oxygen and nitrogen.

We have already seen this happen on Mars. Once home to a thick atmosphere, Mars lost its magnetic field billions of years ago. With no defense, its atmosphere was eroded into space, leaving behind a cold, barren wasteland. If Earth follows the same path, it too would become incapable of supporting complex life.

Cosmic ray penetration

Cosmic rays – high-energy particles from distant supernovae, black holes, and other extreme cosmic events – travel at nearly the speed of light. Today, Earth's magnetic field deflects most of them. But as it weakens, more of these particles will reach the surface.

The result will be higher DNA mutation rates, increased cancer risks, and severe tissue damage in living organisms. Over time, genetic instability could ripple through entire populations, pushing ecosystems toward collapse and making long-term survival far more difficult.

Disruption of the ozone layer

The ozone layer shields life from harmful ultraviolet (UV) radiation. The Earth's magnetic field indirectly helps maintain it by reducing the interaction between charged particles and atmospheric molecules.

If the field weakens, high-energy particles would accelerate ozone depletion, stripping away its protective capacity. Earth would then be bombarded by UV radiation, leading to higher rates of skin cancer, cataracts, and genetic mutations.

Marine ecosystems would be particularly vulnerable. Plankton are the foundation of the oceans' food chain, but they cannot survive extreme UV exposure. Their loss would cascade through the entire marine biosphere.

Ecosystem collapse

On land, species would face the same pressures. Organisms would be forced to adapt rapidly to higher radiation levels or face extinction. If biodiversity collapses, humanity's survival would be in jeopardy too – making the search for alternative habitats beyond Earth more urgent than ever.

Technological impacts

Modern civilization depends on electromagnetic stability. Satellites, GPS, global communications, and power grids all rely on the magnetic field's protection from solar storms.

Without it, geomagnetic disturbances could directly strike our infrastructure, causing catastrophic failures. A single massive solar storm could knock out power grids, cripple telecommunications, and disrupt navigation systems world-wide. That would leave future generations dependent on unstable, vulnerable technology.

3. Supervolcanoes

Beneath Earth's crust, immense geological forces churn in silence, biding their time until they unleash a catastrophe on a scale few events in history can match.

Supervolcanoes are vast underground magma chambers capable of erupting with almost unimaginable force. Though such eruptions are rare – occurring only once every few hundred thousand years – their impact can be global, reshaping the climate, disrupting ecosystems, and even triggering mass extinctions.

A world-darkening eruption

While a typical volcanic eruption is a local disaster, a supervolcano is a planetary one. When it blows, billions of tons of ash and volcanic gases are hurled into the atmosphere, blocking sunlight for months or even years.

Without sunlight, photosynthesis falters or stops entirely. Plant life withers, food chains collapse, and agriculture grinds to a halt. Species that are unable to adapt to prolonged darkness vanish, biodiversity shrinks, and ecosystems weaken.

But the damage doesn't end there. With fewer plants producing oxygen, the level in the atmosphere falls, adding to the strain on life. Meanwhile, the ash clouds force global temperatures to plummet, ushering in a volcanic winter that could push even the hardiest species to the brink.

A history of catastrophe – a future of uncertainty

Earth's geological record bears the scars of past supervolcanoes. Around 74,000 years ago, the Toba eruption in Sumatra, Indonesia is thought to have triggered a volcanic winter severe enough to almost wipe out early human populations. Yellowstone in the United States – perhaps the most famous supervolcano of them all – has erupted multiple times, each event reshaping landscapes and altering global climate patterns.

Scientists monitor these giants closely, but predicting the exact timing of the next eruption remains out of reach. It could be millions of years away – or far sooner.

Either way, the scale of the threat demands preparation: better early-warning systems, deeper research, and strategies to mitigate the fallout. Because, when the next global-scale eruption comes, we will need every advantage we can get.

4. External cosmic events

Earth may feel like a safe, stable home, but the universe is anything but gentle. Immense, unpredictable forces have shaped our planet's history – and they could do so again. From asteroid strikes to the radiation of dying stars, cosmic events could transform Earth's climate and ecosystems in an instant.

Asteroid and comet impacts

Asteroids and comets have struck Earth many times, leaving craters – and extinction events – in their wake.

The Chicxulub impact in Mexico 66 million years ago is the most famous example. The blast released energy equivalent to billions of atomic bombs, igniting firestorms, plunging the planet into global cooling, and triggering ecological collapse. The dinosaurs perished, and mammals – eventually including humans – rose to dominance in their place.

Today, we can track potential asteroid threats, but space remains unpredictable. The odds of a large impact in the next few million years are high enough that we must take it seriously. Developing the ability to deflect hazardous objects will one day become a critical defense.

Nearby supernovae and gamma-ray bursts

Not all dangers come from collisions. Some arrive invisibly as radiation from the deaths of massive stars. Supernovae unleash torrents of high-energy particles that can strip away planetary atmospheres or irradiate their surfaces.

Evidence suggests that two ancient mass extinctions, the Ordovician (445 million years ago) and the late Devonian (372 million years ago), may have been triggered by nearby supernovae. The radiation likely destroyed much of Earth's ozone layer, exposing life to lethal ultraviolet rays and triggering acid rain.

More recently, in 1987, a supernova in the Large Magellanic Cloud lit up our skies from 168,000 light-years away – spectacular, but totally harmless. A similar event within 50–100 light-years, however, would shred our ozone layer and devastate our ecosystems.

One potential future candidate is IK Pegasi, a binary star system just 46 light-years away. Its white dwarf companion could one day explode as a Type Ia supernova – though current estimates place that event roughly 1.9 billion years in the future.

Passing field stars

Simulations show that a passing star could disrupt the orbits of planets in our Solar System. The odds are small, but over the next four billion years, such an encounter could push Earth into an unstable orbit – or send it spiraling closer to the Sun, rendering it uninhabitable.

5. Human impact

Humanity's influence spans only a blink of geological time, yet our activities have already reshaped the planet. Carbon emissions, deforestation, and relentless resource extraction are accelerating environmental changes that might otherwise have taken millennia.

While these shifts will not render Earth uninhabitable overnight, they are amplifying natural processes, pushing ecosystems to their limits, and creating challenges that demand urgent action. We are altering landscapes, eroding biodiversity, and depleting resources at a pace unmatched in Earth's entire history.

Recognizing our impact is essential – not only to preserve our planet's stability, but to safeguard our own survival.

The future of Earth is, in many ways, in *our* hands.

6. The bigger picture

Earth's capacity to support life is finite. The forces shaping its transformation – from climate change to cosmic events – will play out over immense timescales, but their early signs are already visible in today's ecosystems.

So this is a call to action. The choices we make now will determine whether our civilization thrives or falters. We must invest in sustainable environmental strategies, advance our technology, and look beyond Earth for future homes.

Understanding what lies ahead is the first step toward preparing wisely – ensuring that humanity's story continues long after Earth's chapter closes.

The timeline for Earth's declining habitability

250 million years:

- Global warming leads to mass extinctions

600–800 million years:

- Silicate weathering leads to carbon dioxide depletion
- Loss of plant life
- Collapse of ecosystems
- Oxygen decline

1–1.5 billion years:

- Ocean evaporation
- Solar heating
- Runaway greenhouse effect
- Loss of liquid water

1.9 billion years:

- Possible Type Ia supernova (IK Pegasi) leading to gamma ray radiation and atmospheric instability

2–3 billion years:

- Geodynamo weakening leading to magnetic field decay
- Loss of atmospheric protection
- Increased radiation

5-7 billion years:

- Red giant phase of the Sun

- Obliteration of any remaining structures or ecosystems on Earth
- Obliteration of any remaining atmosphere or water
- Possible consumption by the Sun's outer shell

Other potential mass extinction events along the way

- Supervolcano eruptions
- Asteroid strikes
- Nearby gamma-ray bursts

The final stage

By the time the Sun completes its red giant phase and sheds its outer layers, Earth will either be destroyed or reduced to a barren, lifeless rock orbiting a white dwarf.

Long before this ultimate end, our descendants will need to have found alternative habitats beyond Earth to ensure their survival.

So, if remaining on Earth is not a long-term option, the question is not just when *we must leave, but* where *we might go.*

2

Identifying new worlds

Before we build, we must search. Beyond the solar system, billions of planets orbit distant stars – some scorched, some frozen, a few perhaps just right. We scan the sky not only for escape, but for continuity: places where life might take root, where our stories might begin again. Our hunt for new worlds is part science, and part longing – a quiet hope that somewhere out there, a place with Earth-like conditions might exist.

The hunt for exoplanets in habitable zones

Our search for exoplanets – worlds orbiting stars beyond our Solar System – lies at the heart of our quest to understand the cosmos and secure a future beyond Earth. Among the billions of stars in our galaxy, scientists suspect there are other habitable worlds – perhaps even ones that are capable of supporting life as we know it.

At the heart of this search is the study of habitable zones – the regions around stars where conditions are just right for liquid water to exist. Finding planets in these zones requires advanced telescopes, precision instruments, and innovative methods of detection.

Locating an Earth-like planet is one of the most thrilling challenges in modern science. It calls for international collaboration, cutting-edge technology, and a deep understanding of what makes a planet truly habitable.

Each discovery brings us closer to answering profound questions, such as:

- Are we alone in the universe?
- Could humans one day live beyond Earth?

This hunt is more than just scientific curiosity; it's a glimpse into the future of exploration, survival, and our place in the cosmos.

Determining habitability

A planet's position in its star system is important, but its habitability depends on a complex interplay of environmental and physical factors. Understanding these conditions helps scientists identify worlds that may harbor life – or serve as our future homes.

The habitable zone

Often called the "Goldilocks zone," the habitable zone is the orbital sweet spot where temperatures are neither too hot nor too cold for liquid water. Its location depends on the star's size and type: smaller, cooler stars like red dwarfs have close-in habitable zones, while larger, hotter stars push them farther out.

But simply being in the habitable zone is not enough. A planet must also have the right atmospheric and geological conditions to sustain life:

- **Atmospheric composition**

  It needs a dense, stable atmosphere to regulate temperature, shield against radiation, and support biochemical processes. Too little atmosphere (as on Mars) leads to extreme temperature swings, while too much (as on Venus) can trigger a runaway greenhouse effect.

- **Magnetic field strength**

  A strong magnetic field protects against solar wind and cosmic radiation, preventing atmospheric erosion. Without it, a planet's air can slowly vanish – as happened on Mars.

- **Geological activity**

  Volcanism and plate tectonics help regulate climate by cycling gases like carbon dioxide. This keeps temperatures stable over long timescales.

Together, these factors determine whether a planet can sustain not just microbial life, but complex, evolving ecosystems.

Detecting exoplanets

Astronomers use several key techniques to find and study exoplanets.

1. The transit method

One of the most successful approaches, the transit method has revealed thousands of exoplanets through missions such as NASA's *Kepler* and the *James Webb Space Telescope.*

How it works

When a planet passes in front of its star from our perspective, it blocks a tiny fraction of the star's light, causing a dip in brightness. Repeated transits can reveal a planet's size, orbit, and even hints of its atmosphere.

By studying how starlight filters through that atmosphere, scientists can detect gases such as water vapor, carbon dioxide, and methane. These are all potential signs of habitability.

Limitations

Not all planets transit their stars from our viewpoint, and some signals can be mimicked by other phenomena, including starspots (cooler, darker regions on a star's surface). To reduce false positives, astronomers often combine transit data with other detection methods.

> Starspots are known as sunspots in our Solar System.

2. Measuring radial velocity

This method detects the subtle gravitational pull a planet exerts on its star, which cause the star to "wobble" slightly in space.

How it works

As the star moves toward us, its light shifts toward blue; as it moves away, it shifts toward red – an effect known as the Doppler shift. Measuring these changes can reveal a planet's presence, mass, and orbit.

Strengths and limitations

Radial velocity is particularly effective for finding massive planets, such as gas giants, that may not transit their stars. Smaller planets produce weaker signals, which makes them harder to detect. Stellar activity, such as starspots, can also mimic the effect, so this method is often paired with transit observations for confirmation.

3. Direct imaging

For centuries, astronomers dreamed of seeing exoplanets directly. The challenge is that stars outshine their planets by millions or billions of times, making them almost impossible to see.

How it works

Instruments such as coronagraphs and starshades block the star's glare, allowing the planet's faint light to be seen. Infrared imaging is especially useful, as planets emit heat that can be detected beyond visible wavelengths.

What it reveals

Direct imaging can uncover a planet's temperature, atmospheric composition, and even weather patterns. By analyzing the reflected light, scientists can search for biosignatures including oxygen, methane, and carbon dioxide. Tracking a planet's motion over time can also reveal its orbit, and offer clues about its climate and stability.

This method works best for large planets orbiting far from their stars. Smaller, Earth-like worlds remain difficult to spot, but next-generation telescopes – including the *James Webb Space Telescope* and the upcoming *Nancy Grace Roman Space Telescope* – are pushing the boundaries of what is possible.

4. Gravitational microlensing

Among the most intriguing tools in the exoplanet hunter's kit is gravitational microlensing – a phenomenon first predicted by Einstein's theory of general relativity. It allows astronomers to detect planets not by their *own* light, but by the way their gravity bends and magnifies the light of a more distant star.

How microlensing works

When a planet orbits a star, their combined gravity acts like a cosmic lens. As they pass in front of a background star, the light from that star is subtly distorted and amplified. From Earth, telescopes register this as a brief, measurable brightening.

The beauty of microlensing is that it doesn't require us to see the planet directly. Instead, it uses the warping of space-time as a signal. This fleeting alignment can reveal a planet's existence, mass, and orbital characteristics – even if it's far beyond the reach of other detection methods.

What microlensing reveals

This technique excels at finding planets in wide orbits, as well as rogue planets that drift alone through space without a parent star. Unlike transit methods, which depend on a planet crossing its star from our viewpoint, microlensing can detect planets thousands of light-years away.

By studying the shape and duration of a microlensing event, scientists can sometimes infer the presence of multiple planets, moons, and even planetary rings. In theory, it could also reveal artificial megastructures – such as Dyson spheres – should any exist. We'll explore these in more detail in Chapter 5.

Crucially, microlensing can detect planets as small as Earth, which makes it a valuable tool in the search for potentially habitable worlds.

Limitations and challenges

Microlensing depends on rare, precise alignments between distant stars and planetary systems. Telescopes must continuously scan vast areas of the sky to catch these short-lived events. Once an event is over, it can never be repeated or directly followed up, which limits the amount of detail we can gather.

While this method can reveal a planet's mass, distance, and orbit, it tells us nothing about its surface, atmosphere, or whether it harbors water or life. Nevertheless, as our observational technology improves, microlensing will help us map the hidden architecture of planetary systems across the galaxy.

Why finding habitable exoplanets matters

Our search for habitable worlds is more than an exercise in curiosity; it is a quest that could shape the destiny of our species.

- **Backup plans**

  Discovering planets with Earth-like conditions could give us alternative homes if Earth faces environmental collapse or other existential threats. Studying these worlds will lay the groundwork for future migration. However, interstellar travel remains a formidable challenge, as we will explore in Chapter 7.

- **Understanding life**

  Perhaps the most thrilling possibility is finding life beyond Earth. Even the discovery of simple alien microbes would transform our understanding of biology, revealing new evolutionary pathways and expanding our sense of life's diversity.

- **Scientific collaboration**

  The hunt for habitable planets is a global effort that unites astrophysicists, biologists, engineers, and space agencies. This shared mission drives technological innovation, strengthens international partnerships, and pushes the boundaries of discovery.

Each new exoplanet we find is more than a data point; it's a step toward a future where humanity thrives not just on one world, but among many.

What does "habitable" really mean?

To be truly habitable, a planet must offer an environment where humans could thrive for the long term – either in its natural state or with the help of terraforming.

Liquid water is essential, of course, but it is only one piece of the puzzle. True habitability depends on a finely tuned balance of atmospheric, geological, and environmental conditions that remain stable over time.

1. Atmosphere

Composition

A breathable atmosphere needs the right mix of gases – roughly 21% oxygen, 78% nitrogen, and trace amounts of carbon dioxide, argon, and others. Too much oxygen risks hyperoxia and widespread fire hazards; too little and we couldn't survive.

Pressure

For our lungs to function efficiently, atmospheric pressure should be close to Earth's sea-level standard: about 1 atmosphere (101.325 kPa, 760 mmHg, 14.7 psi). Lower pressures can cause hypoxia, while higher ones can lead to respiratory distress.

Protection from radiation

A dense atmosphere is vital for shielding life from harmful cosmic rays and ultraviolet (UV) radiation. Ozone plays a key role here, absorbing UV light and preventing widespread biological damage.

2. Water

Liquid water

Water is the universal solvent for life's chemistry. It regulates temperature, supports agriculture, and makes complex ecosystems possible.

Hydrological cycle

A stable water cycle – with precipitation and evaporation distributing water across the planet – is essential for climate balance.

Accessibility

Freshwater sources such as rivers, lakes, and aquifers must be available for drinking, agriculture, and industry.

3. Temperature

Habitable range

For human comfort and survival, average temperatures should lie between 14°F (-10°C) and 104°F (+40°C). This range supports metabolism and keeps water liquid (or easily meltable).

Seasonal stability

Moderate seasonal changes help to regulate the climate. Extreme shifts – which are common on planets with thin atmospheres, severe axial tilts, or tidal locking – make adaptation difficult.

Heat distribution

Oceans and atmospheric currents also help to spread heat evenly, preventing uninhabitable hot and cold zones.

4. Gravity

The sweet spot for human health lies between 0.8 g and 1.2 g (where 1 g equals Earth's gravity). Lower gravity can cause muscle and bone loss, while higher gravity strains muscles, joints, and the cardiovascular system.

5. Radiation levels

A dense atmosphere, a strong magnetic field, or both are needed to shield the planet against cosmic and solar radiation. Without this protection, the risks of genetic mutations, cancer, and ecosystem damage rise sharply.

Planets orbiting active stars – particularly red dwarfs, which are prone to frequent flares – face additional challenges from bursts of harmful radiation.

6. Light intensity

For plants to thrive, the level of sunlight should be similar to Earth's, ensuring efficient photosynthesis. Too little light weakens ecosystems and stunts growth, while too much risks overheating and radiation damage.

The star's spectrum should include both visible and infrared light to support biological and climate stability.

7. Magnetic field

A strong magnetic field deflects charged particles and prevents atmospheric erosion. As Mars's history shows, losing this shield can strip a planet bare. Magnetic fields usually require a molten metal core, but we could, in theory, engineer an artificial magnetic field. We will look at this in more detail in Chapter 5.

8. Mineral content

Essential minerals – iron, silicon, carbon, and phosphorus – are the building blocks for agriculture, infrastructure, and industry. Nutrient-rich soils that retain water are vital for plant growth.

9. Geological activity

Plate tectonics

Active tectonics recycle gases, regulate climate, and maintain planetary health. Too much seismic activity, however, can make conditions hostile to life.

Volcanism

Moderate volcanism is also beneficial as it replenishes atmospheric gases. However, excessive volcanism can poison the air, block sunlight, and trigger mass extinctions.

10. Orbital and rotational characteristics

Stable orbit

The planet must remain within its star's habitable zone throughout its entire orbit to avoid extreme temperature swings. A circular orbit would create a consistent climate – though this also depends on the planet's axial tilt. An eccentric orbit would create seasons, even if there is little or no axial tilt.

Axial tilt and seasons

An axial tilt similar to Earth's 23.5° or Mars's 25° produces predictable seasons. Extreme tilts create severe climate variations. Low axial tilts, on the other hand, would produce consistent climates and stable temperature zones.

Day length

A rotation close to Earth's 24-hour cycle would support healthy biological rhythms.

> A Martian day is 40 minutes longer than Earth's, and would potentially make human adaptation easier. (See my book *SuperMars* for the full details.)

11. Ecosystem diversity

Microbial life is the foundation of planetary health. It recycles nutrients, enriches soil, and supports plant reproduction through pollinators and other symbiotic relationships.

12. Resource accessibility

Energy sources

Abundant, accessible energy is essential for sustaining industry and infrastructure. This could take the form of solar, geothermal, or nuclear energy, or fossil fuels.

Water and food

Local fresh water and fertile land will reduce the reliance on imports from Earth or elsewhere, strengthening long-term sustainability.

Balancing the criteria for habitability

An Earth-like planet would be ideal, but we may need to accept compromises. Mars, for example, has lower gravity and higher radiation levels, yet we could make it habitable through terraforming and shielding. A high-gravity world might require us to use genetic engineering to adapt our bodies.

The search for habitable exoplanets is as much about human ingenuity as it is about planetary science. Our ability to adapt and innovate will determine whether we truly become an interstellar species.

The major discoveries so far

1. Proxima Centauri b

Just beyond our Solar System lies Proxima Centauri, the closest known star to the Sun. It's a dim red dwarf just 4.24 light-years away. Orbiting it is Proxima Centauri b, the nearest confirmed exoplanet to Earth.

Its position within the habitable zone raises tantalizing questions: could it host life, or even serve as a future human outpost? However, despite its relative proximity in cosmic terms, reaching it would demand massive breakthroughs in interstellar propulsion, and missions would last decades.

A rocky world in a habitable zone

Proxima Centauri b is slightly larger than Earth, with a minimum mass about 1.17 times that of our own planet. This points to a rocky composition. It orbits just 4.5 million miles (7.26 million km) from its star, and completes a "year" in only 11.2 Earth days. That might sound perilously close, but Proxima Centauri is far cooler and dimmer than the Sun, so the planet's surface would not necessarily be scorched purely due to its proximity.

If it has an atmosphere rich in greenhouse gases, climate models suggest it could maintain stable temperatures and even liquid water. But its atmosphere currently remains a mystery; observations made by the *Hubble Space Telescope* have been inconclusive so far.

The threat of stellar flares

Like many red dwarfs, Proxima Centauri is prone to violent stellar flares, which can unleash bursts of intense radiation. Without a strong magnetic field, the planet could lose its atmosphere, leaving its surface dangerously exposed. Whether it has such a magnetic shield is still unknown – though its size and density suggest it should be capable of generating one via the geodynamo process.

Water world or frozen wasteland?

Computer models hint that Proxima Centauri b could be an ocean world or – if it lacks a thick atmosphere – a frozen one. The *James Webb Space Telescope* and other instruments are now searching for atmospheric biosignatures such as oxygen, methane, and water vapor.

Proxima Centauri b might also be tidally locked, with one hemisphere in perpetual daylight and the other in endless night. However, these extremes could be moderated by its oceans or a dense, circulating atmosphere, so habitability could still be a possibility.

The next step: paying it a visit

Its closeness to Earth makes Proxima Centauri b a prime target for future exploration. The proposed *Breakthrough Starshot* mission aims to launch a fleet of 1,000 tiny light-sail probes by around 2036. Traveling at 15–20 percent of light speed, the probes could arrive within 20–30 years, and send back images and data that will reach Earth about four years later.

Whether it proves to be a water-rich haven, a frozen wasteland, or a barren rock, Proxima Centauri b offers us a rare chance to study a potentially habitable world up close – and perhaps take our first real step toward the stars.

2. The TRAPPIST-1 system

Located 40 light-years away, in the constellation Aquarius, the TRAPPIST-1 system has captivated scientists and the public alike.

This ultra-cool dwarf star hosts seven Earth-sized planets, with three of them orbiting in its habitable zone.

A compact and unusual planetary system

All seven planets orbit much closer to their star than Mercury does to the Sun. Yet because TRAPPIST-1 is so cool and dim, its planets could still be temperate.

Their sizes range from 0.76 to 1.13 Earth radii, suggesting compositions from rocky to icy.

The three most promising planets (TRAPPIST-1e, 1f, and 1g) orbit in the habitable zone.

TRAPPIST-1e has a similar density to Earth's, so it could have a stable atmosphere and a rocky surface capable of supporting liquid water.

Orbital resonance

The planets' orbits are locked in a precise gravitational dance known as orbital resonance. This arrangement keeps their paths stable over long timescales. It might also drive tidal heating, influencing volcanic activity, geology, and atmospheric conditions – which are all key factors in habitability.

Atmospheres and water

Observations by the *Hubble Space Telescope* and *James Webb Space Telescope* suggest that some TRAPPIST-1 planets may have thin atmospheres, possibly composed of hydrogen or other light gases.

Computer models indicate that some could also host vast oceans, while others might be icy. Stellar radiation, however, could significantly affect any water on the surface.

The challenge of stellar flares

Like Proxima Centauri b, TRAPPIST-1 emits frequent flares, which could strip away atmospheres over time. Even so, some planets might still retain their atmospheres if they are protected by magnetic fields. So – for now at least – the door to potential habitability remains open.

A living laboratory for exoplanet research

With seven worlds in one compact system, TRAPPIST-1 offers an unparalleled opportunity to compare planetary atmospheres, surfaces, and climates side by side. *James Webb* continues to search for biosignatures, while engineers are already imagining probes that might one day explore these worlds directly.

3. Kepler's groundbreaking discoveries

NASA's *Kepler* mission revolutionized our view of the cosmos. During its nine-year survey, it identified thousands of exoplanets, and revealed that planetary systems are far more common – and diverse – than we once imagined.

Among its most intriguing finds were super-Earths: planets larger than Earth but smaller than the gas giants. These worlds are especially compelling because they may combine stable atmospheres, liquid water, and geologically active surfaces – making them prime candidates in the search for life.

Some super-Earths orbit within their stars' habitable zones, where conditions could allow oceans to form and climates to remain stable for billions of years. Others may be water-rich worlds or rocky planets with thick atmospheres, each offering a different potential pathway for habitability.

4. Challenges in identifying habitable exoplanets

The discovery of exoplanets has transformed our understanding of planetary systems. But confirming whether any of them could truly support life is another matter entirely. Many of the most promising candidates lie tens, hundreds, or even thousands of light-years away – far beyond the reach of direct exploration with today's technology.

Instead, scientists must rely on innovative observational techniques, sophisticated computer models, and creative problem-solving to work around the limits of distance and uncertainty.

Atmospheric clues

An exoplanet's atmosphere is one of the most important indicators of habitability. To study atmospheres across such vast distances, astronomers use spectroscopy, which enables them to analyze how starlight interacts with atmospheric gases.

This method can reveal the presence of water vapor, oxygen, methane, carbon dioxide, and other compounds – some of which may be chemical hints of life.

But spectroscopy has its limits. Signals from distant planets are often faint, and some non-biological processes can mimic biosignatures. Thin, unstable, or eroded atmospheres add further uncertainty.

The influence of host stars

The star a planet orbits is just as important as the planet itself. Stellar flare activity, radiation output, and long-term stability all determine whether or not a world can support life. But until we have precise data on these factors, the question of habitability must remain open.

The need for exploration

Even with the most advanced telescopes, some questions will only be answered by actually going there. Future interstellar probes will take direct measurements of weather patterns, surface conditions, and potential biospheres – data that no amount of remote observation can ever match.

5. The future of exoplanet exploration

Cutting-edge technology is pushing the boundaries of what we can learn about distant worlds.

Breakthroughs in atmospheric analysis

The *James Webb Space Telescope*, launched in 2021, has already transformed exoplanet research. Its advanced infrared instruments can detect the faint chemical fingerprints of planetary atmospheres, including potential biosignatures. Each new observation deepens our understanding of which worlds might be capable of supporting life.

Looking ahead, the European Space Agency's *PLATO* mission – scheduled for launch in 2026 – will focus on finding and studying Earth-like planets in habitable zones. By identifying rocky worlds with Earth-like conditions, PLATO will refine our criteria for habitability and provide us with further key targets for future missions.

But while many of these worlds lie far beyond our reach, our first steps will be closer to home …

3

The first steps outward

Our first destinations will be close to home: Mars as the first stepping-stone, Europa and Titan as icy laboratories for life, Venus as a cautionary tale and engineering challenge, and enclosed orbital habitats as test-beds for self-sustaining life support.

Mars: our first logical stepping stone

Among all the planets in our Solar System, Mars stands out as the most logical stepping-stone toward interplanetary migration and colonization. Its unique combination of attributes – and its relative proximity to Earth – make it the ideal proving ground for our first permanent home beyond our planet.

Accessibility

Mars is close enough to be within the reach of today's spacecraft. At an average distance of about 140 million miles (225 million km), travel times range from six to nine months, depending on planetary alignment. Earth and Mars line up favorably roughly every 26 months, making regular missions feasible with careful planning.

Similarity to Earth

A Martian day, or sol, lasts about 24.6 hours – the closest match to an Earth day of any planet in the Solar System. Its axial tilt of 25.2° is also similar to Earth's 23.5°, producing comparable – though slightly harsher – seasonal variations.

This makes long-term agricultural and habitation planning more straightforward than on worlds with more extreme day lengths, harsher seasons, or no seasons at all.

Mars's gravity is about 38 percent of Earth's, which is far more favorable for human health than the microgravity of space or the Moon's weak pull (16.5 percent of Earth's gravity).

Resources

Mars has significant reserves of water ice, especially at its poles and possibly beneath its surface. Water is essential, not only for drinking, agriculture, and industry, but also for splitting into hydrogen and oxygen to produce fuel and breathable air.

Its thin atmosphere is 95 percent carbon dioxide – which is another potential feedstock for oxygen production via electrolysis or other chemical processes. Martian soil contains minerals including silicon and iron, which could be used for construction and manufacturing.

Scientific value

Mars's landscape tells a story of a much wetter past, making it a prime target in the search for ancient microbial life.

Studying its geology and climate will also deepen our understanding of how rocky planets evolve, and what makes them habitable or hostile.

Logistical and strategic advantages

Settling on Mars would be a vital rehearsal for colonizing more distant worlds, including exoplanets. Its challenges – cold temperatures, thin atmosphere, and high radiation – mirror those we can expect elsewhere.

Mars is also a strong candidate for long-term terraforming. Over centuries, its environment could be altered to create more Earth-like conditions. A permanent settlement on Mars would also serve as a safeguard against existential threats to humanity on Earth – from climate collapse to asteroid impacts.

Existing momentum

NASA's Artemis Moon missions are already developing technologies and strategies that will transfer directly to Mars. The European Space Agency (ESA) and private companies such as SpaceX are also accelerating plans for Martian exploration and settlement.

Mars exploration

Mars exploration is one of the most ambitious, coordinated scientific and engineering efforts in human history. It involves numerous government agencies, private companies, and international partnerships.

Planners are following a multi-phase approach:

- robotic exploration
- human missions

- infrastructure development
- and, eventually, long-term settlement

1. Robotic exploration missions

Robotic missions are already laying the foundation, gathering data and testing technologies for future human crews.

- NASA's *Perseverance* rover (Mars 2020) landed on Mars in 2021, equipped to collect and store soil samples for later return to Earth. Its companion, the *Ingenuity* helicopter, made the first powered flights on another planet, scouting terrain and expanding exploration capabilities.

- The ESA–Roscosmos ExoMars program includes the *Trace Gas Orbiter* (2016), which studies atmospheric gases such as methane that could hint at biological activity. The *Rosalind Franklin* rover will drill deep into the surface in search of ancient life.

- China's *Tianwen-1* mission – consisting of an orbiter, lander, and the *Zhurong* rover – has been studying Mars's surface geology, climate, and subsurface water since 2021.

- India's *Mars Orbiter Mission 2* (MOM-2) will build on the success of *MOM-1*, continuing its atmospheric and geological studies.

- *Mars Sample Return* (MSR) – a joint NASA–ESA mission planned for the late 2020s – will retrieve the samples collected by *Perseverance* and return them to Earth for detailed analysis.

- Japan's *MMX* mission will explore Mars's moon Phobos, collecting samples that should shed light on Mars's formation and history.

2. Human exploration missions

The ultimate goal is to send humans to Mars.

NASA's Artemis-to-Mars roadmap

The Artemis lunar program will serve as a test-bed for Mars systems, including the *Lunar Gateway* – a lunar-orbiting station that will trial life-support, propulsion, and docking systems.

Lessons from lunar landers will inform Mars descent and ascent vehicles. NASA is aiming for a crewed Mars orbital mission in the early 2040s, followed by surface landings.

SpaceX's ambitions

SpaceX plans to use its fully reusable Starship spacecraft to establish a permanent human presence on Mars, starting in the 2030s.

Founder Elon Musk envisions transporting thousands of settlers over several decades, supported by technologies for autonomous landing, in-situ resource utilization, and habitat construction.

Key challenges

- **Radiation**

  Mars's thin atmosphere and lack of a magnetic field leave its surface exposed to cosmic and solar radiation.

Shielding and – possibly – subterranean habitats will be essential.

- **Psychological and physical health**

  The first human missions could last almost three years, including nine months travel each way and sixteen months on the surface. Crews will face isolation, reduced gravity, and limited medical support.

- **Cost and logistics**

  Missions to Mars are hugely expensive and resource-intensive, requiring sustainable funding and efficient payload delivery.

3. Infrastructure development on Mars

Habitat construction

The use of local materials will be critical in reducing supply needs from Earth. Autonomous construction robots and 3D printers will build habitats, with prototypes being tested on the Moon before they are deployed on Mars.

Energy and power generation

Solar power and small modular nuclear reactors are the leading candidates for producing reliable energy. NASA's Kilopower project is currently developing compact reactors to support remote human missions.

4. Terraforming concepts

Planet-wide terraforming is centuries away, but smaller-scale projects could begin much sooner. Concepts include:

- Greenhouses that use Martian soil and sunlight, with controlled temperature and pressure.
- Releasing greenhouse gases to warm the planet and thicken its atmosphere.

We will take a closer look at the techniques of terraforming in Chapter 5.

5. International collaboration

Mars exploration is increasingly global. Agencies such as NASA, the ESA, CNSA (China), ISRO (India), and private companies are aligning their goals and pooling resources. NASA also plans to share the *Lunar Gateway*'s technologies among its partners including the ESA, Japan, and Canada.

The vision for Mars

Landing a human on Mars will be our second "giant leap for mankind" after the first Moon landing in 1969.

Each small step – from robotic scouts to lunar bases to interplanetary habitats – will bring us closer to becoming a multi-planetary species, capable of surviving long after Earth's habitability ends.

Colonizing Europa

Jupiter's fourth-largest Galilean moon, Europa, presents one of the most tantalizing prospects for future colonization. Beneath its gleaming, icy surface lies an immense global ocean that could support the conditions for life. But before we can settle there, we must first overcome some of the harshest environmental challenges in the Solar System, using technologies at the very edge of our capabilities.

Scientific value

Sealed beneath at least 10 to 15 miles (15 to 25 km) of ice, Europa's subsurface ocean could be the most promising location in our Solar System for finding alien life. If microbial – or even more complex – organisms exist in its depths, they could help answer some of our most profound questions:

- Are we alone?
- What are the universal conditions for life?

Colonizing Europa would give us direct access to its ocean, and place us at the forefront of astrobiology. Europa's dynamic environment – which includes cryovolcanic plumes and shifting ice-ocean interactions – offers us a natural laboratory for studying geological and chemical processes unlike anything we can see on Earth. Insights gained here could reshape our understanding of how planets and moons evolve – and perhaps even reveal clues to the origins of life itself.

Resource availability

Europa's ocean is a vast reservoir of liquid water – a resource that could sustain human settlements for countless generations. It could be desalinated for drinking and agriculture, or split into hydrogen and oxygen to provide rocket fuel and breathable air.

Tidal warming from Jupiter's immense gravity might also create geothermal energy sources beneath the ice. Harnessing this could power habitats indefinitely, making long-term survival possible without the need for constant resupply from Earth.

Learning to exploit Europa's resources would not only secure life there, but also provide a blueprint for habitats in other resource-scarce environments across the cosmos.

Advancing deep-space capabilities

To survive on Europa, we will need to develop technologies capable of withstanding extreme cold, shifting ice, and intense radiation from Jupiter. Habitats will need to be buried deep beneath the surface or encased in thick layers of ice for protection. We could also develop advanced magnetic shielding systems – innovations that might later be adapted for interstellar travel or for exoplanets orbiting more volatile stars.

The tools required to reach and explore Europa's ocean – including autonomous drilling systems, and submersible probes – would also pave the way to exploring other icy worlds, including Saturn's moon Enceladus, Neptune's moon Triton, and even water-rich exoplanets.

A strategic gateway to the outer Solar System

A permanent presence on Europa would be more than a single colony; it would be a strategic foothold in the outer Solar System. From here, we could launch missions to other intriguing destinations:

- Jupiter's moon Ganymede, with its magnetic field and possible underground oceans.
- Jupiter's moon Callisto, with its stable environment.
- Saturn's moon Titan, with its methane lakes and dense atmosphere.

In the longer term, Europa could even serve as a staging ground for missions outside our Solar System.

Testing new colonization models

Europa's extreme conditions will push our ingenuity to its limits. We will need to design ice-covered or subterranean habitats that can withstand intense radiation and the shifting ice crust. These designs could become prototypes for colonies on other icy moons and exoplanets.

Robotic and autonomous systems will be indispensable – drilling through the ice, exploring the oceans, and maintaining infrastructure in an environment where human activity is likely to be limited. Colonizing Europa will not just be about surviving; it will be about learning how to thrive in one of the most inhospitable places imaginable.

Ethical exploration

Europa's ocean could harbor alien ecosystems, which makes it one of the most scientifically and ethically sensitive colonization targets in the Solar System. Unlike Mars, which shows no evidence of active ecosystems, Europa's waters might already be alive – or poised for life to emerge.

Colonization must therefore proceed with extreme caution to avoid contaminating its environments. Earth-based microbes, for example, could irreversibly alter any native biosphere. If we prioritize ethical exploration, we must set a precedent for responsible expansion into the cosmos – balancing discovery, human settlement, and the preservation of alien life.

If we succeed, Europa could become a symbol of how we meet the profound responsibilities that come with venturing beyond our home planet.

Colonizing Titan

Saturn's largest moon, Titan, is one of the most captivating candidates for human colonization. It is a world with an unusual blend of Earth-like traits and abundant resources. Its thick atmosphere and vast hydrocarbon reserves make it one of the most promising locations for our expansion into the Solar System. Yet Titan is also far colder than Mars or Europa, and it presents a very different set of opportunities and challenges.

1. Earth-like features

Dense atmosphere

Titan's atmosphere is one of its most remarkable qualities. Composed primarily of nitrogen with a small amount of methane, it exerts a surface pressure roughly 1.5 times that of Earth's. This makes Titan the only moon in the Solar System where we would not need to wear pressurized suits to step outside; only oxygen masks and thermal protection.

The dense atmosphere also acts as a natural shield against solar and cosmic radiation, giving Titan a critical safety advantage over worlds like Europa or Mars.

Surface lakes and seas

Titan is the only known body beyond Earth with stable liquid on its surface – though here, the lakes and seas are not filled with water but with methane and ethane. These hydrocarbon reservoirs form a landscape of winding rivers, shimmering lakes, vast seas, and coastlines eerily reminiscent of those on Earth.

Beyond their visual appeal, these liquids also represent a valuable energy resource. Future settlers could sail on Titan's methane seas and harness them for fuel – combining exploration with practicality.

Low radiation

Thanks to its thick atmosphere, Titan is naturally shielded from harmful radiation. This eliminates the need for deep underground habitats or heavy shielding, and allows for simpler, more cost-effective settlements. This is a rare advantage in extraterrestrial environments.

Weather and methane hydrological cycle

Titan's methane-based hydrological cycle mirrors Earth's water cycle in many striking ways. Methane evaporates, forms clouds, and falls as rain, carving rivers and filling lakes. This creates weather patterns and seasonal changes, making Titan one of the most dynamic moons in the Solar System. Studying this alien climate will deepen our understanding of planetary processes, while highlighting the surprising parallels between Titan and Earth.

Hydrocarbons

Titan's hydrocarbon reserves are immense, and could possibly exceed all of Earth's known oil and natural gas combined. They could power life-support systems, vehicles, rockets, and manufacturing and assembly plants. This makes Titan a potential hub and refueling station for exploring the outer Solar System.

Water ice

Beneath the hydrocarbon-rich surface lies water ice – and possibly a subsurface ocean of liquid water. This combination of hydrocarbons and water makes Titan one of the most resource-rich environments in the entire Solar System.

Potential for scientific exploration

Titan is a natural laboratory for studying prebiotic chemistry – processes that may resemble those that led to life on Earth. The possibility of methane-based life forms adds an additional layer of curiosity. Discovering such profound forms of life would significantly expand our understanding of biology and the adaptability of life in the universe.

2. The specific challenges of colonizing Titan

While Titan offers a compelling vision of a future beyond Earth, its environment presents formidable challenges. These will demand cutting-edge technology, careful planning, and a pioneering spirit.

Distance from Earth

Orbiting Saturn some 870 million miles (1.4 billion km) from Earth, Titan is one of the most distant colonization targets in the Solar System. Current travel times are around six to seven years – far longer than the six-to-nine-month journey to Mars or the two-to-three-year journey to Europa.

The vast distance also means communication delays of around 1 hour 22 minutes each way. Resupply missions would be costly and infrequent, so colonies would need to be almost entirely self-sufficient, and rely heavily on Titan's own resources.

Cold surface temperatures

With an average surface temperature of -290°F (-179°C), Titan is brutally cold. Habitats would require advanced insulation and highly efficient heating systems. Machinery and electronics would need to be specially designed and shielded to function reliably in such conditions. Overcoming these challenges would push our engineering capabilities to the limit.

Hydrocarbon environment

While Titan's methane and ethane lakes could be a boon for fuel production, they are also highly flammable and toxic. Safe extraction and processing would require specialized equipment and strict safety protocols. Developing these technologies would be essential if we are to turn Titan's alien landscape into a thriving hub.

Low gravity

At just 0.14 g (14 percent of Earth's gravity) Titan's weak pull could cause long-term health issues including muscle atrophy, bone loss, and circulatory problems. Colonists might need artificial gravity in their habitats or intensive exercise regimes to maintain their health.

The low gravity would also affect construction and transport – though it would make launching spacecraft far less energy-intensive than it is on Earth.

Atmospheric composition

Titan's nitrogen-methane atmosphere is toxic to humans, and we will need life-support systems to supply breathable air. Its density, on the other hand, provides excellent radiation protection, making it far safer than many other destinations.

Unknown terrain and surface hazards

The surface of Titan is varied and alien – from wind-sculpted dunes to icy mountains and vast hydrocarbon seas. Its dynamic weather, including methane rain and strong winds, would complicate construction and exploration. Rovers and robots would need advanced navigation systems to adapt to these unpredictable conditions. Mapping and mastering Titan's unique terrain would be both a scientific and engineering triumph.

What about Venus?

Venus is often overlooked in discussions about colonization and terraforming. Its proximity to the Sun, crushing atmospheric pressure, searing heat, and clouds of sulfuric acid make it one of the least hospitable places in the Solar System.

Yet beneath these extremes lie unique advantages and intriguing possibilities. With its Earth-like gravity, natural radiation shielding, and the potential for cities floating in the air, Venus could serve as a proving ground for bold terraforming concepts. However, the Sun's gradual brightening would challenge its long-term future.

1. Venus's advantages over other planets

Similar size and gravity to Earth

In terms of its size and mass, Venus is practically Earth's twin, with surface gravity at about 90 percent of Earth's (0.9 g). Settlers would avoid many of the physiological problems associated with low-gravity environments such as the Moon, Mars, or orbital habitats.

Natural radiation protection

While its surface conditions may be extreme, Venus's dense atmosphere provides exceptional shielding from cosmic and solar radiation. It is far superior to Mars's wispy thin air, which offers almost no protection. This makes Venus one of the safest places in the inner Solar System in terms of radiation exposure – especially for habitats located high in the atmosphere.

Potential for floating colonies

At altitudes of 31–37 miles (50–60 km), Venus's atmosphere offers a sweet spot where the temperature and pressure are similar to those on Earth's surface. Here, we could build floating cities – buoyant structures filled with breathable air, which is lighter than the surrounding carbon dioxide.

Colonists would not need heavily pressurized habitats, and the mild temperatures and low radiation would make life far more comfortable than on the surface. The carbon-rich atmosphere could also be tapped for industrial processes and resource production.

2. The long-term risk of solar brightening

The biggest problem with Venus is its proximity to the Sun, which means it will feel the effects of solar luminosity increases sooner than other worlds. Over hundreds of millions of years, this will intensify its already extreme heat.

Nevertheless, this timescale gives us ample opportunity to explore and exploit Venus before it becomes uninhabitable. Future Venusian civilizations could migrate outward, develop advanced cooling systems, or adapt to the changing environment. In the shorter term, Venus could serve as an intermediate colonization target – the perfect place to test large-scale climate engineering and atmospheric modification.

Terraforming solutions to counteract heating

- Space-based reflectors could block or redirect sunlight, cooling the planet.

- Bioengineered microbes could consume carbon dioxide, reducing the greenhouse effect.
- Introducing reflective particles into the upper atmosphere could lower temperatures.

While these ideas remain theoretical, they demonstrate that even the harshest worlds could be transformed with the right technology.

Alternative colonization strategies

If Venus eventually becomes uninhabitable, its colonies could relocate to Mars, the outer planets and moons, or space-based habitats. Far-future technologies might even allow planets to be relocated, moving Venus outward into a cooler orbit. Or we could build Dyson spheres to create new habitable environments.

These speculative ideas – which we will explore further in Chapters 5 – highlight our capacity to adapt and innovate in the face of cosmic change. If we implement them successfully, they will ensure that our species survives long after the Sun's red giant phase.

Terraforming Venus

Venus presents some of the most formidable terraforming challenges in the Solar System. The surface temperature exceeds 842°F (450°C), the atmosphere is almost entirely carbon dioxide, and the pressure is over 90 times that of Earth's. Its runaway greenhouse effect traps heat so effectively that Venus is hotter than Mercury, despite being 31 million miles (50 million km) farther from the Sun.

But with cutting-edge engineering and bold innovation, we might one day reshape Venus into a habitable world – or at least make *parts* of it accessible to human life.

1. Cooling Venus

If our goal is to build settlements on the surface, our first step would be to bring the temperature down to a survivable level. We could achieve this using large-scale interventions that block sunlight, radiate heat away, or remove carbon dioxide from the atmosphere.

Space-based sunshades (orbital mirrors)

Vast reflective structures in orbit around Venus could block or deflect sunlight, allowing the planet to gradually cool over several centuries. Thin-film solar sails could serve this purpose. While the engineering challenge would be immense, the system would be reversible and non-invasive, as it avoids directly altering the surface.

Removing atmospheric heat via thermal radiation

We could seed the atmosphere with reflective aerosols designed to scatter infrared radiation back into space. This would slowly lower surface temperatures – though the dense blanket of carbon dioxide would make progress slow and energy-intensive.

Large-scale carbon dioxide extraction

Addressing the root cause of Venus's excessive heat would mean removing vast amounts of carbon dioxide from the atmosphere.

Massive air-processing stations, or billions of engineered microbes, could convert the carbon dioxide into oxygen or solid carbon. Mining the atmosphere for carbon could both cool the planet and supply raw materials for industry.

The scale of this task would be staggering, of course. It would require the removal of an almost unimaginable volume of gas, and would likely take thousands of years. Yet, if we could achieve it, it would be one of the most direct paths to transforming Venus's climate and making the planet habitable.

2. Atmospheric modification

Even if we managed to cool Venus's surface to a tolerable temperature, its atmosphere would remain overwhelmingly hostile, with more than 96.5 percent carbon dioxide and almost no oxygen. Making it breathable would require an extraordinary program of chemical and biological engineering, combining cutting-edge technology with centuries of planetary management.

Artificial oxygen generation

One possibility is bioengineering. If we released billions of genetically engineered carbon dioxide-converting bacteria into the atmosphere, they could gradually transform the carbon dioxide into oxygen and organic matter.

Another approach could involve vast airborne algae farms. These would be suspended in the upper atmosphere, where they would photosynthesize on an industrial scale.

Alternatively, electrochemical processing could split the carbon dioxide molecules into carbon and oxygen – which would produce breathable air directly.

However, the sheer mass of Venus's atmosphere means that any of these methods would demand sustained effort over several centuries. We would also need to make major advances in energy generation and processing technology.

Hydrogen injection and atmospheric conversion

Another intriguing concept involves injecting vast quantities of hydrogen into the atmosphere. The hydrogen would react with the carbon dioxide to form water vapor and methane, slowly altering the atmospheric composition.

The hydrogen could be harvested from the gas giants (Jupiter or Saturn) and transported to Venus. While this would create water – which would be a critical resource for eventual colonization – the logistical challenge of moving such immense quantities of hydrogen across interplanetary distances would be staggering.

Terraforming via large-scale chemical reactions

A more direct chemical approach would be to use magnesium, calcium, or silicon compounds to bind carbon dioxide into solid carbonates, permanently removing it from the atmosphere. This would require planetary-scale chemical reactors to process Venus's air into more stable, less toxic components.

Like the other proposals, it would be daunting in scope. But it illustrates the scale of ambition we would need to transform Venus into a habitable world.

3. Floating habitats

Given Venus's crushing surface pressure – the equivalent to being more than 0.6 miles (1 km) underwater on Earth –

some scientists have suggested bypassing the surface entirely and building floating habitats in the upper atmosphere.

High-altitude habitats (floating cities)

Between 31 and 37 miles (50 and 60 km) above the surface lies a sweet spot where temperatures range from 68 to 86°F (20–30°C) and the pressure is similar to sea level on Earth.

Helium- or hydrogen-filled aerostat platforms could float permanently in this region, housing self-contained cities in a controlled, Earth-like environment.

These settlements would offer a far more accessible and immediate path to exploration and habitation than attempting to tame the surface and atmosphere.

Cloud-based biospheres

As well as hosting living quarters, Venus's skies could host self-sustaining biospheres for agriculture, oxygen production, and energy generation. Solar-powered farms could grow food in controlled conditions, while bio-domes would maintain breathable air. These airborne ecosystems would allow humans to live and work high above the toxic clouds, supported by advanced engineering and renewable energy.

AI-controlled atmospheric settlements

To keep the floating cities stable in Venus's dynamic atmosphere, artificial intelligence (AI) systems would need to make continuous adjustments for wind patterns, temperature shifts, and chemical changes. These same systems could also optimize energy harvesting, using solar arrays or wind turbines to tap into the dense atmosphere's energy potential.

A second Earth

Terraforming Venus into a fully habitable world would be a monumental challenge that would most likely take millennia. But floating habitats could allow thriving communities to take root in the skies in just decades. In the meantime, long-term projects to cool the planet, alter its atmosphere, and stabilize its climate could continue on the surface.

Enclosed habitats

Para-terraforming – also known as enclosed habitats or "world houses" – offers a more pragmatic alternative to the monumental task of transforming an entire planet.

Instead of re-engineering a planet's atmosphere and climate on a global scale, this approach focuses on creating sealed, controlled environments where settlers can live comfortably, entirely insulated from the hostile conditions outside.

This is a strategy that is achievable in the near term. It is also highly resource-efficient, and adaptable to almost any extraterrestrial setting. It could be a vital stepping-stone on our path to settling alien landscapes.

1. What is para-terraforming?

At its core, para-terraforming means building large, sealed habitats – or networks of smaller, interconnected shelters – that maintain Earth-like conditions inside. They will have breathable air, stable temperatures, and protection from radiation.

These self-contained environments will leave the wider planetary ecosystem untouched, creating localized habitable zones without the need to alter an entire atmosphere or surface.

They are a blend of advanced engineering and resource-conscious design, and scalable from small research outposts to sprawling enclosed cities. And they are adaptable to a wide range of worlds.

2. Key features of para-terraforming

Controlled atmosphere

Inside these sealed habitats, settlers would enjoy a carefully maintained nitrogen-oxygen mix at Earth-like pressure. This would remove the need for bulky spacesuits or oxygen masks. Life-support systems would constantly monitor, filter, and replenish the air supply to ensure long-term habitability.

Temperature regulation

Climate control systems would keep internal conditions stable and comfortable, regardless of the extremes outside – whether that is the searing heat of Venus or the deep cold of Europa or Titan.

Radiation protection

On worlds with high radiation levels – such as the Moon or Mars – habitats could be shielded with thick regolith* walls, engineered metals, or water barriers.

Underground construction in caves or lava tubes would add further protection.

*Regolith is the loose dust and rock on a planet's surface.

Modular, expandable design

Habitats could start small and grow over time. Early modules might be compact and self-contained, later expanding into interconnected domes, tunnels, or even underground cities as populations increase.

3. Locations best suited to para-terraforming

The Moon

Close to Earth and ideal for testing, the Moon's natural caves and lava tubes are ready-made enclosures that could be sealed and converted into shelters, providing protection from radiation and micrometeoroids.

Mars

With its thin atmosphere and limited radiation shielding, Mars is another prime candidate. Habitats could be built from local regolith, with water sourced from subsurface ice for drinking and agriculture.

Europa

Europa presents a challenging but fascinating option. Its subsurface oceans – and possible geothermal energy – could support enclosed habitats near cryovolcanic vents. But habitats would need heavy shielding to withstand Jupiter's intense radiation.

Titan

Titan's thick atmosphere offers some radiation protection, and its methane and ethane reserves could be tapped for energy. But we would still need enclosed habitats for breathable air and warmth.

4. Construction techniques

Inflatable modules

Lightweight, prefabricated structures – like NASA's Bigelow Expandable Activity Module – could be delivered quickly and deployed rapidly. It could provide instant shelter for the first crews.

3D printing

Robotic 3D printers could use local materials such as ice or regolith to build sturdy, radiation-shielded walls. They could also use sintering techniques to fuse locally collected dust into solid structures.

> Sintering involves heating powdered material until its atoms fuse into a solid mass.

Lava tubes

Natural underground channels on the Moon and Mars could be sealed and repurposed, offering excellent protection with minimal construction on the surface.

Geodesic domes

Transparent or semi-transparent domes could admit sunlight while maintaining the air pressure inside. Shielding or special coatings would block harmful radiation.

5. The advantages of para-terraforming

- **Speed**

  Habitats could be built in months or years, rather than the centuries or millennia required for full planetary terraforming.

- **Efficiency**

  It uses fewer resources, often drawing on local materials to reduce costs.

- **Adaptability**

  It can be scaled from small research stations to vast enclosed cities.

- **Protection**

  It shields settlers from volatile events such as Mars's dust storms or Europa's cryovolcanic eruptions.

- **Preservation**

  It leaves native environments intact, and avoids the ethical dilemmas of disrupting alien ecosystems.

6. Key challenges

- As well as radiation, habitats will need to withstand micrometeoroid impacts, seismic activity, and extreme temperatures.
- They will need reliable, high-output energy sources to power life-support, shielding, and climate control systems.
- Closed-loop recycling systems for air, water, and waste must operate flawlessly, as any failure could be catastrophic.
- Scaling up from small outposts to large colonies would add logistical complexity.
- Long-term living in enclosed spaces raises psychological concerns, including isolation and claustrophobia.

7. Para-terraforming as a pathway to full terraforming

These visions reflect our relentless drive to explore, adapt, and innovate. With each breakthrough, we push the boundaries of what is possible, bringing once-uninhabitable worlds closer to becoming thriving civilizations.

The technologies we develop for enclosed habitats – including advanced energy systems, recycling mechanisms, and structural innovations – could later be scaled up for global terraforming.

In the meantime, para-terraforming allows us to explore and settle on new worlds while preserving their native environments.

While these early ventures will shape our future, their success depends not only on technology but on how we work together.

4

International cooperation

In the next chapter we will look at building starships and exotic propulsion systems. But first we need to think about where the money will come from, along with the ethical implications. What rights will settlers on other planets have? What if they declare independence? What if we discover extraterrestrial life – and they have better weapons than us?

No single nation can shoulder these tasks alone. The scale of what lies ahead demands more than ambition; it calls for pooled resources, shared governance, and a collective agreement to treat space as a common heritage rather than a new arena for rivalry. Without such cooperation, our expansion risks echoing the darker chapters of our history, repeating the mistakes of colonialism, but on a cosmic scale.

Social and political challenges within isolated space colonies

Creating thriving communities beyond Earth will be the most profound undertaking in human history. Over time, these off-world and other-world settlements will develop their own rhythms, cultures, and priorities. But with that growth will come complex social and political challenges: inequality, generational divides, mental health pressures, and questions of identity and governance. We will need to meet these challenges with care and foresight if we are to build fair, resilient societies capable of enduring in environments where survival can never be guaranteed.

Social stratification

One of the most pressing dangers is the rise of rigid hierarchies, where unequal access to life's essentials – oxygen, water, food, and shelter – creates imbalances of power.

Such divisions could quickly breed resentment and unrest. The antidote lies in transparency and fairness: egalitarian policies, shared decision-making, and open resource management. Together these will ensure that no single group can dominate the others.

Intergenerational dynamics

As temporary missions give way to more permanent multi-generational settlements, other tensions may arise. The founding generation might cling to tradition and caution. But their children and grandchildren, raised in these new worlds, might push for innovation, reform, and a broader vision of what their society could become.

Disagreements over the use of resources, governance, and cultural identity are inevitable. Bridging these divides will require leadership that is both flexible and inclusive, and education that honors the past while empowering the future. We must encourage dialogue, and governance must be allowed to evolve, so that every generation feels heard and valued.

Psychological well-being

Life in a confined, isolated environment can weigh heavily on the mind. Without Earth's open landscapes, changing seasons, and diverse social settings, colonists may face anxiety, loneliness, and depression – challenges that could deepen as communities grow.

Supporting mental health will mean more than providing medical care; it will mean nurturing strong social bonds, encouraging creative expression, and making space for joy. Shared celebrations, storytelling, and meaningful rituals can create a strong sense of belonging. Virtual recreations of forests, oceans, and mountain vistas could offer a vital psychological escape, restoring a sense of openness and freedom.

Learning from history and imagination

We can learn a great deal from the experiences of seafaring communities and polar research stations.

Fiction, too, provides valuable thought experiments: *Star Trek*'s Federation envisions an inclusive, cooperative future, while *The Expanse* portrays cultures shaped by scarcity and competition. These, and other real and imagined worlds, can inspire us to balance realism with aspiration, and to build communities that are both resilient and hopeful.

The responsibilities of a multi-planetary species

As we step beyond Earth and our technology advances, so too will the weight of our responsibilities. We will face ethical dilemmas that extend far beyond our current experiences.

These will include questions of human rights – and perhaps even the rights of other life forms. We will also need to consider the care and management of alien environments, the governance of new worlds, and the management of scarce resources.

The choices we need to make will be neither simple nor fleeting. Indeed, they will they be extremely difficult, and their consequences will echo across generations.

1. Environmental management and ethical colonization

Space offers us extraordinary opportunities for exploration and settlement, but it also demands a profound moral compass. We must ensure that extraterrestrial ecosystems remain undisturbed, that we approach planetary engineering with care, and that our expansion does not come at the cost of irreversible harm.

Respecting extraterrestrial ecosystems

The discovery of microbial or complex alien life would present urgent ethical questions. Introducing Earth-based microorganisms into alien environments could irreversibly disrupt native biospheres, alter evolutionary paths, or extinguish entire ecosystems before we even understand them.

To prevent such contamination, missions will require strict biological containment protocols. Every lander, rover, and human explorer will need to meet rigorous sterilization standards.

But preservation is not only about protecting life; it is also about safeguarding the integrity of alien landscapes. Even worlds that initially seem barren could hold geological or atmospheric clues to cosmic history. Reckless mining, terraforming, or industrial development could erase these clues before their value is understood.

Responsible resource extraction – including mining

minerals and water ice, harvesting oxygen from regolith, and utilizing materials from asteroids – will be essential for long-term habitation. But they must be balanced against the environmental cost.

Ethical terraforming and climate regulation

Should we reshape other planets to suit our biology, or should we adapt *ourselves* to suit those planets' natural conditions?

Terraforming could make Mars, Venus, or other worlds more hospitable, but altering their atmosphere or geology carries risks – from destabilizing native life to disrupting climate systems in ways we cannot predict.

Human adaptation could be an ethical alternative to this. We could use genetic engineering to create biomechanical enhancements, or enclosed artificial ecosystems, that allow us to thrive without radically altering the host world.

If these solutions prove more ethical and efficient, then future civilizations may choose them rather than large-scale planetary engineering.

We must also prevent exploitation, and for that we will need interstellar environmental laws. These will define planetary protections, regulate space-based industries, and grant "protected status" to certain worlds – much like Earth's national parks.

2. Governance and social responsibility

Ethical governance will form the backbone of interstellar communities. It will define political autonomy, ensure that citizens are treated fairly, and maintain stability across vast distances.

Planetary governance: independence or ruled from Earth?

A central question will be whether colonies remain under Earth's jurisdiction or become self-governing.

Early settlements will undoubtedly depend heavily on Earth's support. But as they grow self-sufficient, political autonomy will almost certainly follow, bringing unique legal systems, trade policies, and cultural norms.

Interplanetary legal frameworks will need to guarantee fair representation, prevent exploitation, and maintain cooperation between worlds.

This could lead to an alliance – perhaps a United Federation of Planets – that ensures diplomatic stability, mutual trade, and the peaceful resolution of conflicts.

Alternatively, planetary communities might fracture into independent state-like entities, each shaped by its own environment and history.

Artificial intelligence will almost certainly play a major role in governance, managing resources, responding to crises, and enforcing policy. But human oversight must remain central to this, to avoid over-reliance on mechanized judgment.

Ethical treatment of space citizens

Fundamental rights and freedoms must be preserved across all planetary communities. Economic systems should guard against inequality and guarantee the fair distribution of essentials – especially in isolated colonies where survival might depend on skills and efficiency rather than on wealth.

Should other-world economies operate as cooperative, resource-sharing models, or adapt capitalist frameworks for interplanetary trade? Whichever model they choose, those who live and work there will need clear protections for their civil liberties, labor rights, and political representation.

Colonies must also guard against authoritarianism, resisting the temptation to reduce or eliminate freedom under the guise of environmental necessity. Governance should adapt to the realities of life in deep space, but without sacrificing ethical principles.

Similarly, trade agreements, scientific exchanges, and conflict-resolution mechanisms will be essential to keep interstellar civilizations connected rather than divided. Diplomacy will be critical in preventing disputes over resources, territory, and cultural differences.

The ultimate question is this:

- Will our descendants form a united space alliance? Or will competitive planetary interests lead to conflict over territory, sovereignty, or technological dominance?

3. Sustainability and the ethical use of resources

Energy and resource management

If space colonies are to thrive, they must manage their energy, resources, and waste. While raw materials will be essential for their survival, extraction should be carried out responsibly to avoid irreparable damage to planets.

Asteroid mining could offer a promising alternative. Many asteroids are rich in water ice, metals, and minerals that could be harvested without disturbing potentially habitable worlds. The asteroid belt alone could sustain space-based industries for thousands of years, making it an ideal resource hub for interstellar economies.

Energy sustainability will depend on infrastructure designed for long-term efficiency. Dyson swarms built around stars could generate near-limitless power, and reduce our dependence on finite planetary reserves. Deep-space colonies with solar-harvesting stations could build thriving energy-based economies without depleting their host worlds.

But not all settlements will enjoy such abundance. Some will rely on closed-loop energy cycles to sustain life support. These habitats will need fully integrated systems where power generation, waste recycling, and atmospheric regulation form continuous feedback loops. Every molecule of water, oxygen, and energy will need to be reused or replenished, eliminating the wasteful patterns of consumption that have plagued Earth.

Waste management and long-term ecological stability

Space settlements should adopt zero-waste policies, repurposing every material for future use. Unlike Earth, where waste can linger for centuries, interstellar habitats cannot afford environmental degradation.

Advanced recycling technologies will be essential: converting biological waste into soil, metals into construction materials, and discarded plastics into industrial components. In these closed systems, nothing will ever be truly lost.

AI-driven biosphere management will monitor atmospheric conditions, waste conversion, and environmental regulation in real time. Intelligent ecological control systems could allow habitats to function like miniature Earths, balancing oxygen, water, and nutrients without exhausting reserves.

Over time, these systems could evolve into fully independent space biomes that operate autonomously, with no need for imported resources.

Interplanetary cooperation will be critical in preventing reckless expansion and uncontrolled industrialization.

Governments, corporations, and independent settlements will need to work in partnership to uphold sustainability standards. They will also need to ensure that all planetary development respects the long-term ecological balance and avoids environmental crises that could destabilize deep-space civilizations.

4. Scientific innovation and knowledge sharing

Advancing scientific discovery for everyone

Our pursuit of knowledge should rise above territorial boundaries and corporate interests. Interstellar research should be a shared human endeavor, so that each discovery benefits everyone rather than being locked away or exploited by a privileged few.

As space colonies grow, so too will the temptation to monopolize key technologies. Breakthroughs in quantum computing, advanced propulsion, and the use of extraterrestrial materials should be made accessible to every planetary community if we are to thrive together.

Open data-sharing agreements could become the lifeblood of this highly cooperative society. This will prevent individual worlds from hoarding knowledge for political leverage or economic gain. In time, we might see the creation of interstellar research alliances – networks dedicated to sharing scientific findings equitably among spacefaring populations.

Artificial intelligence will be central to this process. AI-driven collaboration could accelerate the transfer of knowledge between worlds, streamline research workflows, and automate complex data analysis. AI-powered laboratories equipped with quantum communication systems might coordinate projects across light-years, ensuring that discoveries are swiftly validated, documented, shared, and applied to real-world challenges.

Cosmic exploration with ethical considerations

Another pressing question is how we should approach potential extraterrestrial intelligence. Should we lead with diplomacy, or risk the perils of territorial or ideological conflict? The answer should be clear – because the latter could be catastrophic.

We will need ethical frameworks that emphasize respect, cooperation, and non-interference. These safeguards will prevent reckless encounters that could destabilize – or even obliterate – either human or alien civilizations.

Spacefaring societies will also face dilemmas over cosmic intervention:

- Should they be allowed to redirect asteroids for mining?

- Or would it only be acceptable if the asteroid posed a collision threat?

- What if deflecting an asteroid to save their world inadvertently endangered another?
- Should we alter stellar conditions or manipulate cosmic environments to make planets more habitable?

Future technology will give us extraordinary capabilities, but we must employ ethical oversight to decide when and how such powers are used. The guiding principle should be simple: avoid disrupting the delicate balance of the cosmos.

We must also decide whether planetary discoveries – including evidence of alien life, breakthroughs in physics, or rare deep-space materials – should be openly accessible or controlled only by a select few.

Scientific integrity and transparency should be paramount, ensuring that such knowledge and resources serve all of humanity rather than being locked behind private agendas or monetized for the benefit of a minority.

These early ventures will shape our future, but their success will depend on more than technology. Success will depend on our ability to work together, to share knowledge, protect environments, and build a civilization worthy of the stars.

Part II

Building new worlds

Imagine a valley under an alien sky: a pressurized habitat humming softly, algae turning starlight to oxygen, children drawing constellations they will name themselves.

This is not a conquest; it is caretaking – of ourselves, of our environments … and of infinite possibilities.

5

Engineering survival

Terraforming Mars will be one of the most ambitious projects in human history.

Later, planet-hopping strategies, self-sustaining space stations, and megastructures such as orbital rings and Dyson swarms will expand our living space and resource base.

Terraforming Mars

Transforming Mars into a world that can support human life will be one of the most ambitious scientific challenges ever conceived. While the idea remains speculative – and far beyond our current capabilities – researchers have proposed a range of methods based on what we know about the planet's climate, geology, and atmosphere. Broadly, these fall into three categories: warming the planet, thickening the atmosphere, and introducing a sustainable ecosystem.

1. Warming the planet

With average surface temperatures of around -81°F (-63°C), Mars is bitterly cold. Raising its temperature will be the first step toward habitability.

Proposed approaches include generating greenhouse gases, deploying space mirrors, redirecting asteroids, stimulating volcanic activity, and altering the planet's albedo (reflectivity).

Greenhouse gas emissions

We could build factories to produce heat-trapping gases – such as carbon dioxide, methane, or chlorofluorocarbons – from Mars's soil or atmosphere. Another idea is to trigger the release of frozen carbon dioxide from the polar caps, using nuclear explosions or concentrated sunlight, to start a runaway greenhouse effect.

The challenge is mostly one of scale. Producing enough greenhouse gases to warm the entire planet would require vast energy resources, as well as technology we do not yet possess.

Space mirrors

Enormous orbital mirrors could focus sunlight onto Mars's surface and poles, melting ice and releasing carbon dioxide and water vapor to accelerate warming. But to be effective, these mirrors would need to span hundreds of miles and be made from millions of segments. Such a feat of engineering is far beyond our present abilities.

Asteroid impacts

There are millions of asteroids in the asteroid belt that lies between Mars and Jupiter. Between 1.1 and 1.9 million of them are larger than 0.6 miles (1 km) in diameter.

Redirecting some of them so they strike Mars would deliver massive bursts of heat, and release subsurface water and ice. While this could thicken the atmosphere and warm the surface, the risks are immense: a miscalculation could devastate the planet's surface.

This method would therefore require extremely precise computer modeling and numerous small-scale trials before any full-scale attempt.

Volcanic stimulation

Artificially triggering volcanic eruptions could release immense heat and gases. We could do this by drilling deep into the planet's crust or setting off underground nuclear explosions, for example. However, safely controlling such activity is far beyond our current engineering technology.

Albedo modification

Mars's pale surface reflects much of the sunlight it receives. Spreading darker material across selected areas could increase heat absorption.

Autonomous robots could mine the dark regolith from Mars's moons, Phobos and Deimos, ferry it to Mars, and distribute it over the surface. This could create localized warming, and potentially influence atmospheric circulation.

But the scale, cost, and energy demands would be enormous. Introducing foreign material could also alter the soil's chemistry in unpredictable ways.

The combination effect

Any warming method that melts ice will release carbon dioxide and water vapor, trapping more heat and amplifying the effect. Even if we cannot warm the entire planet, creating localized habitable zones might provide starting points for settlements and agriculture.

2. Thickening the atmosphere

Mars's atmosphere is extremely thin. Its surface pressure is just 0.6 percent of Earth's, which makes it impossible for liquid water to remain stable, and offers little protection from radiation. Two main strategies have been proposed to thicken it and increase the pressure:

- releasing trapped gases
- importing gases from elsewhere

Releasing trapped gases

Vast amounts of carbon dioxide and other gases are locked in Mars's soil, subsurface, and polar ice. We could release these gases and thicken the atmosphere by drilling into the regolith or using heat. However, current estimates suggest that Mars might not have enough accessible carbon dioxide to make a significant difference. This means we might need to look beyond the planet for additional gases.

Importing gases

One idea is to transport nitrogen or carbon dioxide from other celestial bodies – harvesting it from Venus's atmosphere, for example, and delivering it to Mars. This would potentially enable us to terraform two planets at once.

Another idea is to redirect icy bodies from the Kuiper Belt beyond Neptune's orbit and use advanced propulsion technology to guide them safely to Mars.

Both concepts are largely theoretical, and would require thousands of missions and unprecedented energy resources.

3. Introducing a sustainable ecosystem

For Mars to support human life long-term, it would need a self-sustaining biosphere. This means introducing life forms and systems that can actively modify the environment.

Creating oxygen

We could design genetically engineered cyanobacteria or algae to survive in Martian conditions, where they would convert carbon dioxide into oxygen through photosynthesis. However, without human intervention, Mars lacks the liquid water and nutrients needed for large-scale microbial growth.

Artificial oxygen generators are another option: NASA's MOXIE experiment on the *Perseverance* rover has already demonstrated the ability to convert Martian carbon dioxide into oxygen via electrolysis. Scaling this up could supply enough breathable air for colonists – though not enough to terraform the planet.

Creating liquid water

Mars has abundant water ice in its poles and subsurface. Warming specific regions could release liquid water to form rivers and lakes. But this would only work if the atmosphere was thickened enough to prevent the water from freezing or evaporating.

Soil enhancement

Martian regolith lacks the organic material plants need. One solution would be to introduce Earth-like soil microbes or synthetic fertilizers.

However, Mars's soil also contains perchlorates, which are toxic to plants. Perchlorates dissolve in water, so they could be removed by rinsing the soil. Rinsing the entire surface of Mars would be impractical, but treating smaller areas around habitats could make local agriculture possible.

4. Radiation shielding

Mars lacks a global magnetic field, so its surface is exposed to a constant barrage of cosmic rays and solar radiation. It will be essential that we protect future settlers – and any long-term terraforming efforts – from this radiation.

Thickening the atmosphere

A much denser atmosphere would naturally absorb more radiation, and reduce or eliminate the need for additional shielding. Atmospheric engineering will therefore need to form a key part of our long-term plan.

Underground habitats

In the early stages, before any large-scale atmospheric changes take effect, settlements could be built underground or within natural lava tubes. These structures would provide immediate, reliable protection from radiation while terraforming progresses.

An artificial magnetic field

Another idea would be to deploy a massive magnetic field generator between Mars and the Sun to deflect solar wind and shield Mars's surface. While this is theoretically possible, the energy requirements would be staggering.

5. Terraforming timescales

Even with the most optimistic projections, terraforming Mars would be a project measured in centuries or millennia.

- Initial warming could begin within a century if we can meet the technological and energy requirements.
- Thickening the atmosphere and creating stable ecosystems could take several hundred years.
- Fully transforming Mars into an Earth-like environment would likely take thousands of years.

The exact timeline would depend on overcoming the immense scientific and engineering challenges.

Key challenges

- **Energy demands**

  Terraforming would require energy on a scale far beyond today's entire global output.

- **Technological limits**

  All of the proposed methods remain far beyond our current capabilities.

- **Cost**

  Even the initial stages would require unprecedented investment and international cooperation.

- **Ecological ethics**

  If microbial life exists on or beneath Mars's surface, terraforming could destroy native ecosystems.

★ Amazing future possibilities ★

Exotic metamaterial cloaks for habitat construction

Metamaterials are engineered substances with advanced properties that are not found in naturally occurring matter. They are now being developed for electromagnetic, acoustic, and thermal cloaking.

Recent advances include robust mantle cloaks and flexible shape-shifting lattices, where modular, programmable components can assemble into complex, resilient shapes in response to environmental stimuli.

Future space habitats or colonies could use sheets of nested, adaptive metamaterial to shield themselves from radiation, maintain optimal climates, shift shape, and even cloak themselves from detection. A space habitat might, for example, morph from a compact launch configuration into a sprawling, shielded, climate-controlled city, with built-in invisibility or camouflage to avoid hostile observation.

Advanced cloaking could be embedded into probes, sensors, or wearable materials for personal stealth or protection. It would transform military and civilian strategies, both on Earth and in space.

Researchers are currently moving beyond passive cloaking and developing active cloaking – using parity-time symmetric metasurfaces – to create robust, dynamic invisibility shields.

Creating an artificial magnetic field

For long-term colonization and terraforming, we will need to shield Mars from solar wind and radiation. One of the most ambitious – and speculative – solutions is to generate an artificial magnetic field.

1. A planetary-scale magnetic field

Electromagnetic shields in orbit

We could place a network of powerful electromagnetic generators in orbit around Mars, creating a magnetic field that envelops the planet. If we used superconducting materials, the generators would require minimal energy once they were operational.

In theory, they could protect the entire surface of Mars from radiation and prevent atmospheric erosion. But scaling up to a planet-wide field would demand major advances in superconductors, precision orbital placement, and long-term stability.

Magnetic field generator at L1 Lagrange Point

Placing a magnetic field generator at the Sun–Mars L1 point – where the gravitational pull of Mars and the Sun balance – could create a "shadow" of protection by deflecting solar wind before it reaches the planet. However, building and powering a device large enough to shield an entire planet is far beyond our current engineering capabilities.

2. A surface-based magnetic grid

Superconducting cable network

Laying superconducting cables in a grid across the surface of Mars could create localized magnetic fields to protect habitats and farmland. The modular nature of such a system would allow for gradual expansion. The biggest challenge would be maintaining superconductivity, as it would require cryogenic cooling to around -211°F (-135°C) and a reliable power supply from solar or nuclear sources.

Magnetic hemispheres around habitats

We could equip individual dome-like habitats with their own magnetic field generators, creating localized shields strong enough to deflect harmful particles. This approach would be far more achievable than a regional or planet-wide field, but would leave unprotected areas outside the habitats exposed, including agricultural or mining zones.

3. Magnetizing Mars's core

The boldest concept is to restart Mars's natural geodynamo by re-heating its core. Earth's magnetic field is generated by the motion of molten iron and nickel in its outer core, but scientists believe Mars's core solidified billions of years ago.

In theory, massive heat – from nuclear devices, controlled asteroid impacts, or other methods – could re-melt the core and restart convection.

In reality, this is entirely speculative and comes with enormous risks, including seismic instability and uncontrolled heating.

4. Terraforming with plasma shields

Instead of a magnetic field, we could use advanced plasma technologies to deflect solar wind. Plasma fields, generated by powerful electromagnetic systems, would mimic the protective effects of a magnetosphere. However, plasma shielding is still in its infancy, untested on a large scale, and would require vast amounts of energy to sustain.

5. Key challenges

- **Energy**

  Sustaining an artificial magnetic or plasma field would require power far beyond our current capabilities.

- **Materials**

  Components would need to withstand extreme cold and high levels of radiation.

- **Deployment and maintenance**

  Large-scale systems, whether orbital or surface-based, would require precision engineering and long-term reliability.

- **Cost**

  Such a system would be astronomically expensive, and require global collaboration.

While artificial magnetic field generation remains speculative, it is likely to be an essential step toward Mars's long-term habitability.

Advances in room-temperature superconductors, fusion energy, and large-scale space engineering could make these concepts viable in the centuries ahead.

Terraforming through bioengineering

Harnessing life to transform planets

Bioengineering offers a radically different approach to terraforming. It uses life itself as the primary tool for reshaping hostile worlds into habitable environments.

Instead of relying on vast machines or brute-force engineering, this method would deploy genetically modified microbes, plants, and even entire ecosystems to alter planetary conditions.

By leveraging the adaptability and self-regulating nature of living systems, bioengineered terraforming could create sustainable environments that continue to evolve long after the first settlers arrive.

From machinery to biology

Introducing genetically engineered organisms designed for extreme environments could accelerate planetary transformation, potentially reducing the timescale from millennia to centuries. We would cultivate living systems that work in harmony with the planet, rather than maintaining complex mechanical systems. The organisms would then be allowed to adapt, evolve, and expand on their own.

Designing synthetic life for extreme conditions

The heart of this approach lies in creating organisms capable of thriving where no natural Earth life could survive. Scientists would engineer bacteria, fungi, algae, and plants, giving them specific genetic traits to withstand the unique challenges of alien worlds.

For example:

- Engineered microbes could metabolize toxic gases and release breathable oxygen.
- Specially bred algae could capture carbon dioxide and generate organic material to seed future ecosystems.
- Bioengineered fungi could break down rocky soil into fertile ground for plant growth.

Together, these organisms would form the backbone of a self-sustaining ecosystem, while also laying the foundations for breathable air, nutrient-rich soil, and a stable climate.

Automation and rapid deployment

One of bioengineering's greatest strengths is its potential for automation and scalability. These specially bred life forms would be designed to self-replicate and spread autonomously. Once introduced, they could eventually cover vast areas without any further human intervention.

Robotic systems could deploy the first wave of organisms in targeted regions, where they would adapt to the local conditions and begin modifying the environment.

Over time, they would evolve and spread, potentially transforming the entire planet within a hundred years or so.

Adapting to challenges in real time

Unlike mechanical systems, bioengineered life can be modified in real time to respond to unforeseen challenges. If a planet's temperature shifts unexpectedly, new species could be introduced, or existing ones adjusted, to restore the balance. Harmful organisms could be re-engineered to neutralize any ecological risks.

This built-in adaptability makes bioengineered terraforming highly resilient and responsive to changing planetary conditions.

A new era of planetary management

Terraforming through bioengineering is about working *with* ecosystems rather than *against* them. By harnessing the power of life, we could create living worlds that flourish autonomously, long after human intervention ends.

Not only does this approach promote sustainable planetary management; it also reinforces the interconnectedness of life – on Earth and beyond.

This is very much a transformative vision of the future, where living organisms, rather than heavy industry, hold the key to unlocking our destiny.

Tackling planetary challenges with life

- **Thin atmosphere**

  On Mars, genetically modified cyanobacteria could convert carbon dioxide into oxygen, gradually thickening the atmosphere and supporting plant growth. Over time, this could help establish Earth-like conditions.

- **Toxic atmosphere**

  On Venus, bioengineered sulfur-consuming bacteria could break down sulfur compounds and neutralize the dense, acidic clouds to stabilize the atmosphere.

- **Low temperatures**

  On frozen worlds, genetically modified algae could absorb solar radiation, warming the planet while converting carbon dioxide into oxygen.

- **No liquid water**

  Hydrophilic bacteria and engineered fungi could extract moisture from rock, creating small water deposits. Microbial mats could capture trace humidity from the atmosphere, supporting both settlers and expanding ecosystems.

- **High radiation**

  Radiation-resistant, melanin-rich extremophile bacteria could form protective biofilms over colonies and ecosystems. DNA-repair bacteria could reduce mutation rates, and ensure long-term genetic stability.

- **Toxic soil**

  On Mars, fungi could break down the perchlorates and detoxify the soil. Nutrient-capturing biofilms could enhance fertility, enabling Earth-like plants to grow.

Transforming worlds with bioengineering

Bioengineering has the potential to redefine terraforming. By allowing life to do the heavy lifting, planets such as Mars or Venus could be transformed into flourishing ecosystems far faster than with industrial methods.

A revised terraforming timeline

Stage 1 (decades): Microbial oxygen factories

Genetically modified microbes, such as cyanobacteria, begin converting carbon dioxide into oxygen, steadily thickening the atmosphere and trapping heat. This lays the groundwork for broader ecological development.

Stage 2 (centuries): Larger plants and ecosystems

As the atmosphere stabilizes, bioengineered plants, fungi, and algae spread across the surface, purifying the soil, regulating the climate, and enriching the land with nutrients. Early ecosystems evolve into complex habitats that support a growing diversity of life.

Stage 3 (millennia): A fully developed biosphere

The planet reaches ecological maturity, with forests, rivers, and oceans teeming with life adapted to the local conditions.

A stable climate, with breathable air and fertile soil, supports human settlements, advanced plant species, and animals.

This completes the transformation of a formerly barren world into a vibrant, self-sustaining ecosystem.

Applications beyond Mars and Venus

Creating habitable exoplanets

Bioengineered terraforming could carry our ambitions far beyond the boundaries of the Solar System. It offers the possibility of transforming alien exoplanets – worlds of barren rock or ice, or toxic skies – into thriving, life-supporting biospheres.

By harnessing the power of life, we could seed distant planets with engineered organisms designed to reshape their environments, making interstellar colonization not only possible but sustainable. In doing so, we would extend our presence – and our species' legacy – across the galaxy and beyond.

Restoring Earth's ecosystems

These same techniques could also help heal our own world. Bioengineering could become a powerful tool for reversing the damage caused by human activity and climate change.

- Engineered bacteria and fungi could capture atmospheric carbon dioxide, helping to stabilize the climate.

- Drought-resistant plants could revive arid landscapes and restore biodiversity.

- Specialized microbes could clean up oil spills, detoxify polluted water, and rehabilitate degraded soils.

- Enhanced crops could improve yields in regions suffering from nutrient loss.

These techniques could repair and rejuvenate Earth's ecosystems, ensuring that our planet continues to thrive as we expand into space.

Sustaining space habitats

Of course, bioengineering does not have to be limited to planetary surfaces. It could also sustain life in orbiting space stations, lunar bases, and long-duration spacecraft.

Genetically modified microbes could produce oxygen, recycle waste, and generate food, allowing habitats to operate independently of Earth. Algae and engineered plants could maintain air quality while providing renewable nourishment. Together, these would form the backbone of a closed-loop ecosystem capable of supporting human life indefinitely.

How AI and robotics could assist

Transforming alien worlds into habitable environments is an extraordinary challenge. But artificial intelligence (AI) and robotics could make bioengineered terraforming exponentially faster, safer, and more adaptable.

AI-driven environmental monitoring and analysis

Before terraforming begins, AI systems could map planetary conditions in detail to identify the most effective strategies for transformation.

Advanced climate-simulation models would predict how atmospheric and ecological changes might unfold, guiding the deployment of bioengineered organisms.

Autonomous robots could scan and analyze the soil, pinpointing the perfect locations for synthetic life to take root.

For example, AI-powered probes could map Mars's regolith and locate toxic compounds that need to be neutralized before oxygen-producing microbes can be introduced. AI models could anticipate ecological risks such as runaway greenhouse effects or unstable water cycles, and adjust strategies to keep the terraforming process safe and controlled.

Synthetic biology and AI-guided genetic engineering

AI-assisted genetic engineering could create highly specialized organisms tailored to extreme planetary conditions. Machine learning and advanced DNA synthesis tools could optimize microbes, algae, and fungi for tasks such as oxygen production, greenhouse gas regulation, and nutrient enrichment.

For example, AI could refine the genetic code of cyanobacteria to maximize their efficiency in converting carbon dioxide into oxygen. This would accelerate atmospheric thickening on Mars. AI simulations could also predict how synthetic organisms might evolve, minimizing the risks while ensuring long-term ecosystem stability.

Autonomous terraforming robots

AI-guided robots would be the primary agents of deployment, ecosystem maintenance, and surface modification during terraforming.

Self-replicating biofactories could produce oxygen-generating bacteria, soil-enriching fungi, and radiation-resistant microbes in situ. Robotic seeders might then scatter cyanobacteria across Mars, while drones powered by bioreactors could release algae to stabilize water cycles.

Underground excavators could prepare shielded habitats for microbial colonies – and eventual human colonies – in high- radiation zones.

By reducing the reliance on human labor and ensuring continuous progress, AI systems would make terraforming scalable, adaptable, and significantly faster than traditional methods.

Nanobot-assisted terraforming

Nanobots – microscopic, self-replicating machines – could operate at the atomic level, enriching soil, repairing radiation damage, and regulating atmospheric chemistry.

- On Mars, they could break down iron oxide in the regolith to create fertile soil.
- On Venus, they could work alongside bioengineered bacteria to neutralize the sulfuric acid clouds, reducing toxicity and cooling the atmosphere.

Nanobot networks could also act as self-healing systems, repairing environmental damage and stabilizing ecosystems. But safeguards would be essential to prevent unintended changes, runaway replication, or ecological disruption.

AI-based climate regulation

Over many centuries, AI could manage planetary climates and stabilize emerging ecosystems.

Automated ground-based and orbital monitoring stations would detect early signs of imbalance, and adjust temperatures and atmospheric chemistry in real time.

For example, on Venus, AI-guided greenhouse gas regulators could work alongside cloud-seeding robots to establish stable rain cycles.

Orbital climate control systems

AI-managed satellites and reflectors could fine-tune planetary conditions:

- On Mars, solar reflectors could boost surface temperatures and accelerate ecosystem growth.
- On Venus, massive orbital mirrors could block sunlight, cooling the planet while bioengineered bacteria neutralize the greenhouse gases.
- In deep-space habitats, climate-control satellites could maintain artificial environments for isolated ecosystems.

Long-term oversight

Once the initial terraforming stages were complete, AI would monitor and manage the newly established ecosystems, ensuring their stability and preventing them from collapsing.

Self-learning models could monitor biodiversity and introduce new species as conditions evolve. On low-gravity worlds, robotic systems might maintain the oceans and prevent them from evaporating.

AI systems could even coordinate the management of multiple planets at once – turning interstellar colonization into a synchronized, large-scale endeavor.

Planet-hopping

Imagine a grand cosmic odyssey where we move from one planet to the next – and eventually from one solar system to the next – in search of survival, knowledge, and discovery. In this vision of our future, we transcend the boundaries of a single planet, and build intergenerational civilizations that thrive across multiple worlds.

Turning this vision into reality will demand a great deal more than just imagination, however. We will need to adapt to extreme environments and changing conditions, master new technologies, and overcome challenges that span the practical, the philosophical, and the ethical.

1. The motivation for planet-hopping

Stellar lifespans

Stars have finite lives. Over billions of years, they exhaust their hydrogen fuel, transform, and sometimes die violently. Planet-hopping will allow us to adapt to these cycles by migrating from one solar system to another as conditions change. This will ensure our survival across the vast timescales of cosmic evolution.

Planetary degradation

Even the most promising terraformed worlds will eventually falter. Natural stellar cycles, atmospheric collapse, the loss of magnetic shielding, or the depletion of vital resources might force their inhabitants to seek new homes.

Unlike Earth, which benefits from a natural magnetic field and a relatively stable climate, terraformed planets might require constant maintenance and upkeep – and that maintenance could fail. For example, a meteorite strike or some other catastrophic event could overwhelm the life-support systems and render the entire world uninhabitable.

Just like Earth, no single planet can be a permanent refuge. Our long-term survival demands that we remain mobile.

Celestial dynamics

Planets exist in ever-changing environments, shaped by stellar evolution, asteroid impacts, and orbital shifts. A nearby star swelling into a red giant could render an entire solar system uninhabitable, forcing its inhabitants to migrate – or perish. Mastering planet-hopping would enable us to stay ahead of these cosmic upheavals.

Human expansion

As our population grows and our technology advances, our resource needs (and our curiosity) will outpace what a single planet – or even a single solar system – can provide. Each new settlement will become a stepping stone toward greater horizons.

2. The role of autonomous drones

Our future will depend on fleets of autonomous drones, which will act as explorers, engineers, and guardians rolled into one. These machines will scout out new worlds, prepare them for our arrival, and maintain the infrastructure once we settle there. They will form the all-important advance guard of our interstellar migration.

3. The planet-hopping timeline

Short-term hopping: within our Solar System

Our first steps into space will be close to home: the Moon, Mars, Venus, Europa, and Titan.

The Moon could serve as a staging ground for the first interplanetary missions, while Mars might host the first permanent off-Earth settlements.

These early "hops" will give us invaluable experience in extraterrestrial living and logistics.

Medium-term hopping: to neighboring star systems

As our technology advances, we should be able to reach nearby stars such as Proxima Centauri and Alpha Centauri.

Interstellar probes – like those envisioned in the highly anticipated *Breakthrough Starshot* project of 2036 – will scout those stars for habitable worlds.

Generational ships or cryosleep vessels might carry the first human settlers across the light-years, reaching their new homes within a few centuries.

Long-term hopping: across the galaxy

In the distant future, advanced propulsion – including warp drives, antimatter engines, or traversable wormholes – could carry us across vast swathes of the Milky Way, and even to other galaxies. Megastructures like Dyson spheres might supply the energy for entire interstellar fleets, enabling us to leap from star to star on a galactic scale.

4. Key challenges

- **Resource management**

  Constant relocation will require ultra-efficient systems for transporting supplies and people, generating energy, and recycling resources.

- **Time scales**

  Journeys could last centuries or millennia.
  Maintaining societal continuity and purpose over such massive spans will be a challenge in itself.
  We will explore this further in Chapter 7.

- **Ethical considerations**

  Colonizing and reshaping alien worlds raises questions about our right to alter extraterrestrial ecosystems.

- **Technological complexity**

  Building fleets of intelligent, self-replicating drones capable of operating across interstellar distances will demand unprecedented innovation and careful management, with little margin for error.

5. A vision for solar system-hopping

Picture a cosmic caravan: fleets of space ships, habitats, and drones migrating between stars. Among them are seed ships that carry genetic material, fabricator drones that will construct infrastructure, and mobile colonies that will house entire generations. This is a vision that blends our engineering brilliance with our deep-rooted urge to explore the unknown.

6. The advantages of this approach

- **Survival insurance**

  Spreading ourselves across multiple worlds and star systems will protect our species and its descendants from extinction caused by cosmic disasters or planetary decline.

- **Cosmic exploration**

  We are – and will remain – an exploring species, constantly pushing the boundaries of knowledge.

- **Technological advances**

  The innovations required – from interstellar propulsion to terraforming barren planets – will drive progress across countless fields of science, and benefit life both on and beyond Earth.

7. The ultimate goal

In the far distant future, planet-hopping could evolve into galaxy-hopping. As the Milky Way changes over billions of years, we might then become an intergalactic civilization. This would mark the pinnacle of human adaptation, ingenuity, and ambition, where we thrive in the infinite expanse of the cosmos.

Constructing self-sustaining space stations

Self-sustaining space stations are autonomous habitats, designed to recycle resources and support human life for decades or even longer, without resupply from Earth.

Building these stations will demand careful attention to design, life-support systems, and energy solutions, as well as major technological advances.

Engineering for independence

Designing a space station is one of the most complex engineering challenges of our age. These orbital habitats must shield their inhabitants from the hazards of space while enabling research, innovation, and perhaps even permanent human settlement far from Earth.

Every design choice must balance resilience, adaptability, and cutting-edge technology to ensure safety and functionality in a freezing, radioactive vacuum.

Radiation shielding

Radiation is one of the greatest threats to long-term space habitation. To protect crews, station walls could incorporate layers of shielding materials such as polyethylene, which is lightweight yet effective at absorbing radiation, or water, which can double as both protection and storage.

In high-radiation environments, more advanced solutions might be needed. For example, the space station could generate its own magnetic field to deflect charged particles.

Modular architecture

Modern space stations are incredible feats of modular engineering. Rather than being built in one piece, they are assembled in stages, with interlocking modules designed for specific purposes: laboratories, living quarters, greenhouses, manufacturing bays, and so on.

This modularity allows for gradual expansion, easy upgrades and replacement, and customization for deep-space missions.

The *International Space Station* (ISS) is a prime example of a modular space station. It's an international collaboration that links modules from the USA, Russia, Japan, and Europe into a single, unified habitat.

This approach ensures resilience, redundancy, and the ability to evolve with new technologies.

Artificial gravity

Microgravity takes a huge toll on the human body. It causes muscle atrophy, bone loss, fluid redistribution, and other issues. Future long-term space stations will likely incorporate systems to simulate gravity.

Rotational habitats – where sections of the station spin to generate centrifugal force – can mimic Earth-like gravity if they are rotated at the right speed. Toroidal (doughnut-shaped) stations are a natural fit for this design, as they offer a continuous ring of habitable space with consistent artificial gravity.

Construction techniques

1. In-Situ Resource Utilization (ISRU)

Using local materials from celestial bodies will reduce the need to launch vast quantities from Earth. Lunar regolith, for example, could be 3D-printed or sintered into solid blocks for building habitats, shielding, and infrastructure. We could also mine asteroids for metals such as iron, nickel, and aluminum, which would enable us to fabricate structural components, tools, and machinery where they are needed.

2. Robotic assembly

Space is a harsh and dangerous workplace for humans. Autonomous robots equipped with artificial intelligence and 3D printers could fabricate and assemble space station components directly in orbit.

NASA's SPIDER project is an early example of this. It's designed to build trusses, antennas, and other structures from delivered or locally fabricated materials.

In the future, swarms of specialized robots could construct entire space stations autonomously. They will work in orbit continuously, adapting to challenges in real time, and achieving levels of precision that would be impossible for human crews alone.

3. Prefabrication and launch

Today's space habitats begin as prefabricated modules that are built and tested on Earth, then launched into orbit for assembly. Reusable rockets from companies like SpaceX are making this approach more affordable, as well as allow-ing larger and more complex modules to be delivered and connected into fully operational stations.

Sustainability and longevity

For a space station to be truly self-sustaining, it should be designed for closed-loop operation, self-repair, and total reliability and resilience in emergencies.

1. Closed-loop systems

Every molecule of air, water, and organic matter must be recycled. Waste management systems will convert organic waste into fertilizer for hydroponic or aeroponic farms, which will in turn provide food and psychological benefits.

Air will be continually scrubbed of carbon dioxide and replenished with oxygen. The oxygen is most likely to come from the electrolysis of recycled water. Multi-stage purification systems will reclaim water from condensation, hygiene stations, and waste streams.

2. Self-repair and maintenance

Automated maintenance systems will detect, diagnose, and repair problems before they escalate. Self-healing materials – such as polymers embedded with micro-capsules of repair agents – will seal micrometeoroid punctures automatically.

AI-guided repair drones will handle more complex tasks such as replacing solar panels or repairing structural damage, minimizing – or eliminating – the need for human spacewalks.

3. Resilience to emergencies

Space stations must always be prepared for the unexpected. Redundant life-support, power, and communication systems will be essential to ensure everyone survives any failures. Backup batteries, fuel cells, and alternative energy sources will provide critical power redundancy. Emergency shelters or escape pods will offer safe havens or evacuation routes. Robust communications will keep the station connected to Earth or other spacecraft, even in cases of crisis.

Self-sustaining space stations

Self-sustaining space stations are the stepping stones to our future among the stars. They will serve as platforms for exploration, hubs for development, and symbols of our expansion into the cosmos.

Starting with lunar outposts, they will develop into interstellar staging grounds, offering us a glimpse of what a truly spacefaring civilization might look like.

The *Lunar Gateway*

The *Lunar Gateway* is a bold, multinational project led by NASA in partnership with Europe's ESA, Japan's JAXA, and the Canadian Space Agency. Designed to orbit the Moon, this modular station will act as a staging ground for crewed lunar landings and, eventually, missions to Mars.

Its architecture blends cutting-edge innovations with lessons learned from the *International Space Station*. It will be strategically positioned as a prototype for sustainable operations far from Earth, allowing astronauts to transfer between Earth, the *Gateway*, and the lunar surface.

This in turn will enable deeper exploration of the Moon's terrain and resources. It will also serve as a test-bed for life-support systems, robotics, and advanced energy solutions – technologies that will be essential for more distant missions.

Deep-space habitats

Placing self-sustaining stations in deep space – in the asteroid belt or near the gas giants, for example – would unlock huge opportunities for exploration and economic growth.

These habitats could become hubs for asteroid mining, extracting water ice, metals, minerals, and hydrocarbons to support interplanetary industry. This would minimize the need for costly launches from Earth, and make missions far more efficient and sustainable.

Deep-space stations could serve as logistics hubs, resupply points, research stations, and refueling depots, extending our reach further into the Solar System.

The stations' proximity to celestial bodies such as Europa, Titan, or Ceres would also offer unprecedented opportunities for scientific study, from probing alien oceans to analyzing exotic atmospheres.

Interstellar space stations

While today's space stations focus on near-Earth operations, their ultimate potential lies in interstellar exploration. In the far future, they could evolve into generational ships, sustaining communities of thousands of people for centuries as they travel to habitable exoplanets.

Many researchers conclude that a generational ship would need a population of at least 10,000 to ensure genetic diversity. But there is considerable disagreement. Some argue that only 100 or so would be necessary, while others call for as many as 44,000.

Other space stations could act as stepping stones between stars. These would be strategically positioned waypoints for fleets carrying settlers, supplies, and scientific equipment.

Powering such stations could involve advanced fusion reactors, or even energy harvested directly from stars via Dyson sphere fragments.

These concepts would transform space stations from orbital platforms into dynamic tools that power our spread across the galaxy.

Ethical and social considerations

As we push further into space, building self-sustaining stations could raise profound ethical and social questions. These projects will need to align with our values of well-being, environmental responsibility, and global cooperation.

Environmental impact

Space stations must avoid adding to the growing problem of orbital debris (more commonly known as "space junk"). Every discarded fragment increases the risk of collisions, and threatens both current and future missions.

Engineers are developing construction methods that minimize waste. They are also exploring ways to recycle or safely de-orbit obsolete craft and materials.

We must ensure that stations – particularly those near other worlds – do not contaminate pristine environments or disrupt potential alien ecosystems. Outposts on the Moon, Mars, and other celestial bodies will need to follow strict planetary protection protocols to preserve their natural states.

Global cooperation

International collaboration will be essential. It's not just a matter of sharing the immense costs. By working together, nations will ensure equitable access to space, and avoid monopolization or exclusion. Shared governance frameworks could keep space as a domain for peaceful exploration and collective discovery.

In this way, the construction of space stations could become a symbol of what humanity can achieve when it transcends geopolitical divisions in pursuit of a common goal.

Psychological challenges

Life on board a space station will test the human mind as much as the body. Long-term isolation, confinement, and separation from Earth could lead to stress, anxiety, and depression.

Early explorers will be carefully screened and selected for their mental resilience. But as we move forward, station designs will need to prioritize mental and emotional health.

For example, recreational areas for exercise, relaxation, and creative pursuits will be vital. Natural simulations – such as virtual forests, oceans, and open skies – could help maintain a connection to Earth. Thoughtful interior design, with warm lighting, soothing colors, and personalized spaces, will make habitats feel more human-centered.

Astronauts and future settlers should then flourish in their new – and extraordinary – environments.

Artificial space colonies and megastructures

Artificial space colonies and megastructures will form a central pillar of our long-term survival strategy – our "grand plan" to endure long after Earth becomes uninhabitable. These self-contained habitats and monumental constructs will be capable of sustaining vast human populations in the depths of space.

They will be engineering marvels, but they will also be much more than that. They will stand as declarations of our resilience, creativity, and determination to explore uncharted frontiers.

Artificial space colonies will offer us a practical yet revolutionary answer to the challenges of interplanetary living. Designed to recreate Earth-like conditions, they will provide breathable air, fresh water, food production systems, and gravity. Rotational habitats – such as O'Neill cylinders and the Stanford Torus – will use centrifugal force to generate gravity, preserving long-term physiological health. Advances in hydroponics, aeroponics, and closed-loop life-support will turn these colonies into self-sustaining ecosystems, capable of supporting human life indefinitely.

Megastructures, by contrast, will push our cosmic ambitions to dazzling extremes. Vast constructs such as Dyson spheres and stellar engines could harness the power of stars to provide near-limitless energy. These would represent the next stage of human growth, enabling interstellar travel, terraforming distant worlds, and creating habitats that could house billions of people.

O'Neill cylinders

First proposed by American physicist Gerard K. O'Neill in his 1976 book *The High Frontier: Human Colonies in Space*, O'Neill cylinders are one of the most ambitious and ingenious visions for space colonization. Designed to host millions of people in self-sustaining environments beyond Earth, they represent a massive leap in both imagination and engineering.

Structural design and construction

Each O'Neill cylinder is a colossal tube: 3.7 miles (6 km) in diameter and 3–20 miles (5–32 km) long. Two counter-rotating cylinders are connected by a central hub. Their rotation creates artificial gravity via centrifugal force, while their counter-rotation cancels out any gyroscopic effects and ensures they remain stable.

The structure would be built from lightweight, high-strength materials such as titanium, aluminum, or carbon nanotubes – ideally sourced from asteroids or the Moon. Their outer shells would be engineered to withstand micro-meteoroid impacts, cosmic radiation, and solar flares – potentially using thick regolith layers, water shielding, or magnetic fields.

Interior design and social structure

Inside the cylinders, alternating strips of land and windows would create zones for cities, farmland, and recreation. Landscapes would mimic Earth's, with forests, lakes, and urban areas interspersed with agricultural zones. In larger cylinders, there might even be areas of wilderness. All of this would be designed to support community, creativity, and a vibrant cultural life.

Hydroponics and aeroponics would ensure food self-sufficiency, while biodiverse ecosystems would maintain soil health and protein sources.

Giant angled mirrors outside the cylinders would direct sunlight through the windows, filtering out harmful UV radiation and creating a natural day-night cycle. The internal atmosphere would match Earth's, and would be maintained by plant-based recycling systems and carbon dioxide scrubbers.

Sustainability and life support

Solar panels on the outer shell would power the life-support systems, agriculture, and industry. Closed-loop water systems would recycle every drop, supplemented by reserves mined from asteroids. Agricultural zones would provide fresh food, eliminating the need for resupply from Earth.

Feasibility

Building megastructures like O'Neill cylinders would require major breakthroughs in robotics, autonomous construction, and nanotechnology. Mining and mass-driver systems would be needed to extract and deliver materials from the Moon or asteroids. Costs would run into the trillions of dollars, demanding decades of international collaboration.

Long-term benefits

Despite their immense cost, O'Neill cylinders could provide permanent human settlements in space, relieve population pressures on Earth, and preserve its environment.

Space-based manufacturing in microgravity could produce materials that are impossible to create on Earth. And relocating certain industries off-planet could also reduce the ecological strain.

The Stanford Torus

First proposed in 1975 during a study by the NASA Ames Research Center, the Stanford Torus is a smaller, more manageable alternative to the O'Neill cylinder. It's a bold concept that is now edging closer to feasibility due to recent advances in materials, automation, and energy systems.

Overview

The Stanford Torus would orbit at geostationary altitude – 22,236 miles (35,786 km) above Earth's equator. It would serve as both a habitat and a launchpad for deep-space exploration. Its compact size makes it more achievable with near-future technology, and it could house thousands of residents in a sustainable, Earth-like environment.

Design

The Torus is a ring-shaped structure, approximately 1 mile (1.6 km) in diameter. It rotates once per minute to generate artificial gravity. Its interior would be divided into zones for living, agriculture, work, and recreation. Angled mirrors would direct sunlight inside, simulating day-night cycles and supporting plant growth.

Advantages

As it is significantly smaller than an O'Neill cylinder, the Stanford Torus requires far fewer resources and less infrastructure. This would make it a practical first step in large-scale space colonization. Its combination of gravity, fresh food systems, and green spaces would create a sense of normalcy for its residents.

Key challenges

- Radiation shielding would remain a major hurdle.
- Its smaller scale would limit population capacity, making it less suitable for long-term mass migration. But as a prototype, it could pave the way for larger habitats.

Bernal spheres

First proposed in 1929 by Irish scientist John Desmond Bernal, the Bernal sphere was one of the earliest serious visions for a permanent human habitat in space.

Positioned in orbit around Earth or other celestial bodies, the rotating spherical habitats would be entirely self-sufficient, with integrated food production and closed-loop life-support systems.

Bernal envisioned the spheres as sustainable homes for hundreds or thousands of residents. Each sphere would provide a compact yet complete environment for research and industry – or even permanent settlement.

Design

A Bernal sphere would measure about 0.6 miles (1 km) in diameter – small enough to be practical, yet large enough to feel spacious. It would rotate on its axis to create Earth-like gravity on its inner surface.

Inside, the habitat would be divided into residential areas, public infrastructure, and agricultural sectors, just like the other megastructures.

The outer shell would double as a protective barrier and a light-management system, with angled mirrors directing sunlight into the interior and maintaining a natural day-night rhythm.

Advantages

- **Compact efficiency**

  The spherical shape maximizes usable space and minimizes material requirements, making it ideal for smaller populations, research stations, or industrial outposts.

- **Radiation shielding**

  Its geometry makes it easier to shield than elongated tubular or cylindrical designs. Layers of water or regolith could protect inhabitants from cosmic rays and solar radiation.

- **Versatility**

  Smaller and simpler than the other vast megastructures, a Bernal sphere could serve as our practical first step toward permanent life beyond Earth.

Key challenges

Even at this scale, a Bernal sphere would require major advances in engineering, in-situ resource utilization, and life-support technology.

Its limited size would restrict population size and growth, making it best suited to specialist roles rather than mass migration.

Constructing the protective shell and interior infrastructure would require enormous quantities of raw materials – probably sourced from the Moon or asteroids. However, the mining, processing, and transport systems needed to do this remain beyond our current capabilities.

Dyson spheres

Proposed by British-American physicist Freeman Dyson in the 1960s, the Dyson sphere is not a habitat but a bold energy-harvesting concept designed to capture the entire energy output of a star. Achieving this would elevate humanity to a Kardashev Type II civilization.

In popular culture, Dyson spheres are often depicted as solid shells encasing a star. However, Dyson's original idea was far more practical: a Dyson swarm – a vast constellation of solar-collecting satellites orbiting in complex patterns. This decentralized network would avoid the almost-impossible engineering of a single rigid shell. Instead, the satellites would absorb stellar energy and beam it wirelessly to where it was needed: planets, space stations, or interstellar outposts.

Purpose and potential

We would not build such a structure around our own Sun, of course, but around other stable stars without habitable planets. A fully realized Dyson swarm would eliminate energy scarcity and would fuel amazing new technologies on a scale we can barely begin to imagine.

With so much power available to them, interstellar ships could reach relativistic speeds (10 percent of light speed or faster), distant worlds could be terraformed, and planetary eco-systems could be maintained indefinitely.

Challenges

Even a simplified Dyson swarm would require astronomical amounts of raw materials – far beyond what Earth alone can supply. We would need to mine entire moons, planets, or asteroid belts, and create vast interplanetary manufacturing and logistics networks.

Other obstacles include:

- **Heat dissipation**

  Capturing most of a star's energy would generate immense heat, which we would then need to radiate away to prevent system failure.

- **Structural resilience**

  Solar collecting satellites would also need to endure intense radiation, gravitational forces, and constant stellar activity including flares.

- **Material science**

  Components would have to be built from advanced, ultra-durable materials capable of surviving for centuries in extreme conditions.

Inspiration

While they are far beyond our current technological reach, Dyson spheres and Dyson swarms are powerful symbols of what might become possible in the future. They challenge us to think on a stellar scale, and to imagine a future where the stars themselves are the engines of our destiny.

Let's go even bigger: ultra-large space habitats

Imagine a space settlement so vast that it could easily accommodate the entire global population – and many more.

Ultra-large space habitats represent the next evolutionary tier of megastructures, surpassing even O'Neill cylinders in their scale, ambition, and technological sophistication.

Capable of housing billions or even trillions of people, they would solve planetary overcrowding and resource scarcity, while also acting as launch points for interstellar exploration.

These immense, interconnected habitats could replicate – and even exceed – Earth's ecological diversity and societal complexity. They would function as multi-purpose civilizations in their own right, transforming the Solar System, and eventually the galaxy, into a vibrant network of thriving communities.

McKendree cylinders

First proposed by NASA engineer Tom McKendree, these colossal structures are vastly upscaled versions of O'Neill cylinders. Each one is designed to house tens of billions of people in a fully self-sufficient environment orbiting Earth or another celestial body.

With their vast interior spaces, McKendree cylinders could function as complete self-contained worlds.

Design

Each cylinder could stretch up to 62 miles (100 km) in length and 12–24 miles (20–40 km) in diameter. To withstand the immense stresses of rotation and the internal atmospheric pressure, construction would require carbon nanotubes or materials of comparable tensile strength – hundreds of times stronger than steel, yet much lighter.

Key challenges

The sheer volume of material needed would far exceed Earth's capacity to supply it. Once again, entire asteroids, moons, or other celestial bodies would need to be mined, which would require revolutionary advances in autonomous mining and orbital manufacturing. Metal-rich asteroids in the asteroid belt would likely serve as primary resource hubs.

Maintaining the cylinders' physical and ecological stability on such a scale would be equally daunting. Rotational forces would need precise tuning to ensure uniform gravity. At the same time, balancing the ecosystems, air quality, water cycles, and food production for billions of people would require advanced closed-loop systems and constant AI-driven monitoring.

Earth's primary backup

With its capacity to support tens of billions of residents, a McKendree cylinder could serve as a primary refuge if Earth became uninhabitable.

Bishop rings

First conceived in 1997 by American engineer and author Forrest Bishop, the Bishop ring is a massively scaled-up version of the Stanford Torus. With a monumental diameter of 1,240 miles (2,000 km) and 310 miles (500 km) wide, it too could house billions of people.

Its inner surface would cover 1.2 million square miles (3 million square km) – equivalent to the size of India or Argentina. It could host dense cities alongside preserved wilderness, creating a natural balance between human activity and environmental harmony.

Its atmosphere would be contained within its 200-mile (322-km) high walls, and retained by gravity alone. This means that the ring would be open to space and would not need a roof.

One of the Bishop ring's biggest strengths is its material efficiency. It open outer shell significantly reduces resource requirements while maintaining structural stability and environmental sustainability.

Dyson swarm habitats

We have already explored the concept of Dyson swarms as stellar-scale energy collectors. Dyson swarm habitats take this idea a stage further by integrating millions of individual space habitats into the swarm. This would transform an energy megastructure into a massive, decentralized, star-orbiting civilization.

Overview

Each of the millions of habitats – which could be mega-structure such as O'Neill cylinders, Stanford Tori, or some other design – would orbit the star independently. But together they would form a vast, interconnected society powered directly by the star at the center of the swarm.

Design

Collectively, these habitats could host trillions of people. The swarm's modular nature allows it to grow incrementally over millennia, adapting to technological advances and the needs of the population. Each habitat would be fully self-sufficient, with direct access to virtually unlimited solar energy.

Advantages

- **Modular growth**

  Habitats can be added gradually, reducing the upfront costs and allowing flexibility.

- **Resilience**

  There is no single point of failure. Damage to one habitat will not compromise the rest of the network.

Key challenges

As with the other megastructures, we would need to acquire vast amounts of resources, build the habitats in orbit, and keep them ecologically sustainable with trillions of inhabitants.

But this design comes with two further hurdles: connectivity and coordination.

- **Transportation networks**

  High-speed transit systems, such as autonomous space shuttles or magnetic-propulsion craft, would be needed to bridge the vast distances between the habitats.

- **AI-driven management**

  Millions of habitats would require intelligent systems to keep the swarm synchronized, stable, and efficient, while monitoring their energy and resource use.

★ Amazing future possibilities ★

Dyson swarm ecological circuits

Recent climate modeling has revealed some profound, unintended ecological consequences of Dyson swarms. The models found that even a partial swarm could drastically alter planetary climates by redirecting, blocking, or re-radiating solar energy. This is why we would not build Dyson spheres or Dyson swarms around stars with habitable planets.

Far-future civilizations might need to develop Dyson swarm ecological circuits. These would be dynamic, actively-managed satellite constellations designed to optimize energy harvesting while sustaining – or modifying – planetary or stellar-scale ecologies.

This could include orbital solar array management systems that regulate planetary temperature, control photosynthesis spectra, or deliberately engineer climates across multiple worlds by redistributing starlight.

These systems might be coupled to AI-controlled biosphere governors, planetary-scale environmental sensors, and programmable orbital geometries.

In some speculative models, Dyson swarm ecological circuits could also serve as signals of civilization in the search for extraterrestrial intelligence (SETI). This is because the energy re-radiated by the swarms would produce distinct infrared technosignatures.

However, the energy ambitions of advanced civilizations might collide with ecological limits. Advanced ethical frameworks would then be needed to balance energy extraction with planetary habitability and long-term cosmic sustainability.

Shell Worlds

Imagine an entire planet sealed beneath a colossal engineered shell, which stabilizes the planet's atmosphere and climate. This is the concept of a Shell World – one of the most audacious ideas in speculative engineering.

Unlike traditional terraforming, which attempts to alter a planet's existing environment, a Shell World adds an external layer that traps a breathable atmosphere, regulates temperature, and shields against radiation. In doing so, it could transform even the most hostile worlds – from tiny moons to planets the size of Mars – into habitable spaces where civilizations could live, flourish, and expand.

The shell could be constructed from ultra-light, ultra-strong materials such as graphene or advanced carbon composites. Beneath it, an Earth-like artificial atmosphere would be maintained. Integrated solar mirrors and reflectors would control light and heat, ensuring a stable climate.

For Mars, with its thin atmosphere and lack of a magnetic field, a Shell World could trap oxygen, stabilize temperatures, and create a pressurized environment, with the shell also providing protection against radiation and the vacuum of space.

Similar strategies could be used on icy moons like Europa or Enceladus, taking advantage of their natural water ice.

Advantages

- The shell would be built directly on the planet's surface, making efficient use of local resources. It might be supported by giant legs, poles or towers.

- It would give us complete control of the climate, including custom weather systems.

- It would protect the planet from cosmic radiation and micrometeoroids.

- The host planet or moon would provide gravitational stability, making it practical for large populations and industry.

Challenges

- Encapsulating a planet the size of Mars would require unimaginable quantities of material, and technologies we have yet to develop. Mining entire asteroids or moons would be necessary, and we would need to develop sophisticated autonomous mining, processing, and transport systems.

- The shell would need to withstand impacts, solar and cosmic radiation – and the immense stresses of its own mass.

- Life-support systems on this scale would have to recycle air, water, and nutrients with unprecedented efficiency. AI-driven monitoring would be essential to prevent environmental imbalances.

Niven rings (Ringworlds)

First imagined by the American science fiction author Larry Niven in his 1970 novel *Ringworld*, a Niven ring is a megastructure of staggering proportions. It would form a continuous band around a star, offering an almost unlimited surface area for habitation.

With a radius of 1 astronomical unit (93 million miles or 148 million km) – the distance of Earth from the Sun – its inner surface could host natural landscapes, ecosystems, and cities on a scale millions of times greater than Earth's.

It would rotate to generate artificial gravity.

"Shadow squares" – vast panels orbiting the ring – would regulate sunlight and create day-night cycles.

Advantages

- Direct access to the star's energy for billions of years.
- Virtually unlimited living space, eliminating overpopulation and resource scarcity.
- Modular construction, allowing incremental building and expansion.

Challenges

The raw materials required to build a Niven ring would be far beyond the capacity of a single planet. Building it would likely require the dismantling of multiple moons, planets, or entire asteroid belts. Stability would be a critical engineering challenge too, as even a slight misalignment could cause catastrophic wobbling or collisions.

Matrioshka Brains

A Matrioshka Brain combines the stellar-scale energy capture of a Dyson sphere with the computational power of a superintelligence.

Named after the Russian nesting dolls, it would consist of multiple concentric shells around a star, with each layer harvesting energy and using it to power vast computing platforms.

Unlike a Dyson sphere, which focuses solely on energy collection, a Matrioshka Brain would channel that energy directly into computation. It would potentially be powerful enough to simulate entire universes, host digitized human consciousness, and solve the deepest scientific and philosophical questions.

Design

- The inner layers surround the star and capture its energy for computation.
- The outer layers absorb waste heat from the inner shells and convert it to energy for computation, maximizing efficiency.
- It could also integrate physical habitats alongside the digital environments.

Potential

Such a system could host fully immersive simulated worlds, allowing uploaded human minds to live in limitless digital realities. Freed from biological and physical constraints, lifespans could be indefinite, creativity boundless, and exploration unconstrained by distance.

> **★ Amazing future possibilities ★**
>
> **Matrioshka brain governance models**
>
> Civilizations that reach this Type II+ status will need to design governance frameworks as intricate as the megastructures themselves. Governance models explored in speculative literature and research forums suggest that the best approach might be massively distributed swarm-like rule.

This would be composed of AI node-neurons, each with delegated territories, functions, or ethical constraints. The nodes would coordinate transparently across the system.

Governance could involve liquid democracy*, massive parallel simulation of policy outcomes, or even the incorporation of constitutional physics within the computational substrate. This means that core ethical and operational rules would be hardwired into the system's infrastructure.

This could be extended to cosmic regulation. In this case, decision nodes would vote not only on internal affairs, but on interstellar treaties, contact protocols, and the ethical treatment of "computational citizens" – including uploaded minds, digital animals, or evolved algorithmic life forms.

*Liquid democracy is a combination of direct democracy and representational democracy. A person can either vote directly on an issue or transfer their vote to a delegate who votes on their behalf. The delegate might amass votes from numerous people and vote on behalf of all of them, or further delegate the votes to someone else.

Stellar engines and migrations

Even after leaving Earth, we might face new cosmic threats. For example, we might encounter wandering black holes or supernovae. Stellar engines could offer a solution.

These megastructures are designed to move entire solar systems to safer regions of the galaxy.

Shkadov Thruster

The Shkadov Thruster is one of the most frequently discussed stellar engine concepts. It consists of a gigantic reflective sail positioned on one side of a star. The sail reflects the star's radiation asymmetrically to create a continuous, gentle thrust. Over a period of several millennia, this could alter the star's trajectory, also carrying its planets and habitats along with it.

Solar shell migration

Another idea is solar shell migration. This is the coordinated movement of numerous interconnected habitats around multiple stars, which "hop" across the galaxy over eons. For example, networks of Dyson swarm habitats could share resources and adjust their orbits collectively, powered by the stars themselves.

Challenges

Once again, the energy demands would be immense, but they could be met by stellar-scale energy capture, such as from a Dyson swarm.

Robotic self-replication could allow construction at the necessary pace and scale. This would make interstellar migration (and our long-term survival) a potential reality.

Once we've shaped these environments, we must power and sustain them – and that means pushing the boundaries of science.

6

Scientific innovation

From re-engineering Earth's orbit to shielding against supernovae to managing ecosystems in closed loops, our survival will depend on bold, precise engineering. Advances in astrophysics, propulsion, and AI-driven design will make deep-space habitation both possible and sustainable.

Harvesting energy from space

As we push deeper into the Solar System – and eventually beyond it – our ability to extract and use energy directly from space will be essential. Self-sufficient colonies, spacecraft, and orbital habitats will depend on advanced energy solutions to power their life-support systems, industry, and exploration. Mastering these technologies will be the key to becoming a truly spacefaring civilization. It will also unlock new frontiers in innovation, survival, and discovery.

1. Mining resources for energy

Helium-3 mining

Helium-3 is a rare isotope that is ideal for nuclear fusion. It offers the promise of immense power with minimal radioactive waste. It's found in lunar regolith and in the atmospheres of the gas giants such as Jupiter and Saturn.

On the Moon, billions of years of solar wind have embedded helium-3 into the surface dust. Robotic mining and refining systems could extract it and transport it to fusion reactors aboard space stations, creating a virtually limitless energy source.

Mining helium-3 from gas giants, however, remains far beyond our current capabilities. Their crushing gravity, turbulent atmospheres, and extreme conditions make this a challenge for future generations of advanced spacecraft and robotics.

Asteroid mining for hydrocarbons and metals

Asteroids are rich in resources that could be used for energy production and infrastructure.

- Water-rich asteroids contain hydrogen and oxygen, which can be separated via electrolysis to create rocket fuel or sustain life-support systems. This would enable spacecraft to refuel in space rather than relying on costly launches from Earth.

- Carbonaceous asteroids contain organic compounds that could be used to generate energy.

- Metallic asteroids are abundant in nickel, iron, and other metals that are ideal for building and maintaining power infrastructure in space habitats.

These asteroids could become important interplanetary refueling stations and supply depots, and would support both exploration and industry.

Uranium and thorium extraction

Some asteroids and planetary crusts may contain uranium or thorium. These are ideal fuels for nuclear fission reactors.

They could be mined and refined to power long-term colonies on the Moon, Mars, or in the asteroid belt.

Nuclear fission's reliability and high output make it invaluable for deep-space habitats, especially where solar energy is scarce. Unlike the other concepts we have discussed, nuclear fission is already a proven technology that supplies around 10 percent of Earth's electricity.

2. Tapping into stellar energy

Harnessing the energy of stars and other celestial phenomena could transform our ability to live in space.

A. Solar energy: direct harvesting

The Sun is our most abundant and dependable energy source. Space-based solar arrays could capture its power without the limitations of Earth's atmosphere. This will avoid interruptions from day-night cycles, unreliable weather, or atmospheric scattering. Vast orbital farms of photovoltaic panels could supply electricity directly to spacecraft, space stations, and lunar or Martian colonies.

Another concept is wireless power transmission, where mirrors or satellites beam solar energy via microwaves or lasers to receiving stations on Earth or other worlds. We will need to overcome challenges in beam safety and efficiency, but such a system could form the backbone of an interplanetary power grid.

Dyson swarms

In the far future, Dyson swarms could power everything from self-sufficient space stations to entire interplanetary civilizations. Building the swarms would require engineering on an unprecedented scale, but it would mark a new era in the evolution of life and technology across the galaxy.

B. Gas giant atmospheres: electromagnetic energy

The gas giants are immense sources of electromagnetic activity. Jupiter's magnetic field is the largest in the Solar System, and it generates intense radiation that we could capture using specialized orbital equipment. We could then convert it into usable power for nearby space stations.

Hydrogen mining

These planets' atmospheres are also rich in hydrogen – a key fuel for fusion reactors. Robotic systems might one day be able to harvest and refine it, turning the gas giants into hubs for energy production. There are formidable challenges to overcome, of course, but the rewards would be immense.

3. Harnessing gravitational and kinetic energy

Tidal forces

Gravitational interactions between planets and their moons create tidal stresses that generate heat beneath the surface. A prime example is Europa's subsurface ocean, which remains liquid under the icy crust due to Jupiter's gravity. Underwater turbines could capture the tidal motion, while thermal systems could convert tidal heat into electricity for habitats both above and below the ice.

Orbital energy capture

Electrodynamic tethers are long, conductive cables that interact with a planet's magnetosphere, generating electricity as they move through the magnetic field. This technology could power space stations which orbit planets with strong magnetospheres – including Earth, Jupiter, and Saturn.

Tethers could also adjust a space station's orbit without the need for traditional propulsion.

Asteroid spin energy

Many asteroids rotate rapidly, and are a potential source of kinetic energy. Specialized tether systems or turbines could harvest some of this energy without affecting the asteroid's orbit. Mining stations could also use it to power extraction equipment, life-support, and communications systems – making them completely self-reliant.

If we combined all of the approaches we have looked at so far in this chapter, we could create a diverse, resilient energy network to power our expansion across the Solar System and beyond.

4. Exotic energy sources

Our search for innovative energy solutions is driving us to consider the most extraordinary – and currently untapped – reserves of power the universe might hold. Exotic energy sources such as antimatter, black holes, dark matter, and zero-point energy could one day revolutionize space station design, interstellar propulsion, and our civilization's sustainability among the stars.

A. Antimatter

Antimatter is the most energy-dense fuel known. It's capable of producing vastly more power per unit mass than any conventional energy source. Today, scientists can synthesize antimatter in particle accelerators, where high-energy collisions create antiparticles in minute quantities. In the future, we might be able to collect it from natural cosmic sources – perhaps near black holes or through interactions with cosmic rays.

If we could harness it, antimatter could power spacecraft capable of traveling at near-light speeds. This could open the door to practical interstellar journeys. However, antimatter's immense promise is matched by its tremendous challenges. It is prohibitively expensive to produce, extremely difficult to contain, and it annihilates instantly if it comes into contact with normal matter. Safe storage would require sophisticated magnetic traps within high-vacuum chambers.

B. Black hole energy

Hawking radiation

Theoretical physics suggests that microscopic black holes emit tiny amounts of radiation as they slowly evaporate – a process known as Hawking radiation. If we could learn to create and control these black holes, their decay could provide an almost limitless energy source.

Accretion disks

Rotating black holes generate immense amounts of energy in their accretion disks – the superheated gas and debris that spirals around their event horizons.

In theory, spacecraft or space stations could capture some of this energy. The Penrose mechanism even describes how we might extract energy from a black hole's rotation.

> A black hole's event horizon is the boundary where its escape velocity exceeds the speed of light.
>
> Anything that crosses this boundary will remain inside it forever, as nothing can escape – including light.
> This is why black holes are black.

C. Dark matter

Dark matter remains one of the great mysteries of modern science. Invisible and interacting only weakly with normal matter and light, it is thought to make up about 85 percent of the universe's total matter.

If we could learn to detect it, manipulate it, and harness it, dark matter could become a revolutionary energy source for space stations and propulsion systems. While this idea is still entirely speculative, the sheer abundance of dark matter makes it an enticing target for future research.

D. Zero-point energy

Quantum physics points to the existence of zero-point energy – subtle fluctuations in space-time that persist even in a perfect vacuum. In theory, these quantum fluctuations could contain vast amounts of energy.

If we could master the technology to tap into them, we would gain access to a virtually infinite power supply.

★ Amazing future possibilities ★

Dark matter microbial ecosystems

While the true nature of dark matter remains elusive, some speculative biologists have suggested that microbial dark matter could exist. These forms of life might be based on unknown physics or chemistry, and use dark matter particles. If they exist, they would be completely undetectable with our current methods.

Future detection might rely on their indirect effects, unique isotope ratios, gravitational anomalies, or non-electromagnetic interactions.

The existence of dark-matter ecosystems would fundamentally redefine our understanding of biology, cosmology, and the diversity of life-matter relationships.

In classical biology, "microbial dark matter" refers to an immense array of uncultured and functionally mysterious microorganisms – most of which have only been detected via metagenomics*. This is not the same as organisms composed entirely of dark matter, which is what we are discussing here.

If structured dark matter can form stable, complex systems, then dark matter biospheres (or "shadow biospheres") could exist throughout the universe. They might even live adjacent to our own biospheres, perhaps influencing or parasitizing familiar life forms in ways we do not understand.

*Metagenomics is the sequencing of DNA or RNA from an entire community of microorganisms from a specific environment, rather than sequencing an individual specimen.

For example, scientists might take a bulk sample from soil, water or the human gut to gain a comprehensive overall view of the microbial life in that environment.

5. Advancing energy efficiency and transmission

As we venture deeper into the cosmos, two innovations could redefine how we generate and distribute power in space: energy beaming and self-sustaining systems.

Energy beaming

This concept uses wireless transmission – via microwaves or lasers – to send power across vast distances. These systems could provide a constant flow of renewable energy, enabling cleaner, lighter, and more sust-ainable missions.

Self-sustaining systems

Closed-loop reactors recycle waste heat and spent fuel into reusable energy, creating highly efficient power cycles. On-board fusion systems, fueled by hydrogen or helium-3, could push sustainability even further, turning space stations into energy-independent hubs capable of supporting their own needs. They could even share their surplus power with other outposts.

6. Key challenges

Our dream of powering self-sustaining space stations with energy harvested from celestial bodies is as ambitious as it is inspiring – but the obstacles are formidable.

Technological constraints

Gas giant planets and black holes are extreme environments: crushing pressures, violent storms, immense gravity, and intense radiation. Operating in these conditions will require huge breakthroughs in robotics, materials science, and energy transmission.

AI-guided machines will need to function autonomously under these stresses. We will also need to develop new alloys and composites that can endure cosmic wear and tear. Efficient wireless energy transmission systems will be essential if we are to move the energy from such hostile environments to where it is needed.

Cost and feasibility

The economic challenges are as great as the technical ones. Launching, maintaining, and scaling this kind of infrastructure will demand an enormous amount of investment, as well as unprecedented international cooperation.

Not only will we need breakthroughs in propulsion and orbital construction, but also in resource-sharing if we are to make our incredible ambitions achievable.

Ethical considerations

Mining or altering celestial bodies for energy risks disturbing their natural states and destroying potential ecosystems. Even apparently barren environments could hold immense scientific value as untouched records of cosmic history.

If we discover extraterrestrial life, the moral stakes rise sharply. Do we prioritize our survival, or respect the right of other life forms to exist undisturbed?

We will need to establish robust ethical frameworks to balance our ambitions against our responsibility to preserve the wonders of the universe.

Moving Earth into a wider orbit

As the Sun ages, its luminosity will steadily increase, bathing Earth in more and more solar radiation and making our planet progressively less hospitable. One radical proposal to counter this is not to abandon Earth for Mars or artificial habitats, but to move the entire Earth–Moon system into a wider orbit – perhaps a similar orbit to Mars's. This would reduce the amount of sunlight Earth receives, meaning that we could maintain a stable climate and potentially extend the planet's habitability for billions of years.

While the concept is theoretically sound, the technical and energy demands would be staggering. Current estimates suggest it could take a billion years before we develop the capability to move Earth effectively. However, the Sun's relentless brightening could render our planet uninhabitable within just a few hundred million years. If we are to attempt such a monumental feat before we run out of time, research and innovation will need to accelerate dramatically.

Moving Earth and the Moon

Gravity assists

One method borrows from a technique we already use in spaceflight: the gravitational "slingshot." By redirecting asteroids along precisely calculated paths, their gravitational pull could nudge Earth's orbit outward in tiny increments. Over millions of interactions – spread over millions of years – the cumulative effect could be significant.

However, the precision and patience required would be almost unimaginable. A single miscalculation could cause a catastrophic impact or destabilize Earth's rotation.

Solar sails or laser propulsion

Imagine a reflective sail thousands of miles wide, tethered to Earth, catching the Sun's radiation to exert a continuous outward push. The radiation could be supplemented by powerful laser arrays, which would deliver targeted beams of force.

The challenge is that solar radiation pressure is minuscule compared to Earth's gravitational binding energy. The sails would need to be vast, and the laser arrays unimaginably powerful. We would need major breakthroughs in materials science and energy generation to achieve this.

Controlled nuclear or fusion explosions

A more controversial approach involves detonating nuclear fission or fusion devices in space near Earth to impart outward momentum. While this is theoretically possible, the risks are severe. It could destabilize Earth's tectonic plates, damage the atmosphere, or trigger unintended orbital changes. Millions of precisely controlled detonations would be needed over geological timescales, which makes this approach both logistically and politically daunting.

Artificial gravity tractors

Another concept suggests placing massive engineered objects known as "gravity tractors" in space near Earth to exert a steady gravitational pull. This approach would require exotic, ultra-dense materials and colossal amounts of energy to slowly shift Earth's orbit. In theory, it would offer a more stable and controllable solution, but the technology remains speculative.

The consequences of moving Earth

Relocating Earth would have profound implications for the planet, the Moon, and the delicate balance of our climate systems.

The impact on the Moon

The Moon stabilizes Earth's axial tilt, ensuring predictable seasons. If Earth's orbit changed, the Moon's orbit would need to be adjusted in tandem to preserve its stabilizing effect. Without it, Earth's tilt could vary wildly, triggering extreme and unpredictable climate shifts. Tidal patterns – which are vital to coastal ecosystems and infrastructure – could also be disrupted.

Biological and climate effects

The outward shift would need to exactly match the Sun's growing brightness. If we move too quickly, Earth could cool down – to the point where current ecosystems collapse. We might need some terraforming measures – such as orbital solar reflectors or atmospheric engineering – to fine-tune the climate. Any misstep could profoundly alter agriculture, water cycles, and biodiversity.

The risk of orbital instability

Gravity-assist maneuvers, in particular, could cause orbital chaos. Even a small mistake in the transfer of momentum could send Earth into an unstable path – or, in the worst case scenario, eject it from the Solar System entirely.

We would need ultra-precise control systems, advanced computer modeling, and multiple layers of redundancy to avoid such outcomes.

And, of course, moving a planet is as much an ethical challenge as it is a technological one. The stakes are nothing less than the survival – or potential annihilation – of all life on Earth.

Solar shading

In the event that we are unable to move Earth into a wider orbit before the planet becomes uninhabitable, we must consider other strategies. One of the most promising is solar shading. This involves deploying enormous structures in space to reflect or block a portion of the incoming sunlight.

Solar shading could stabilize the climate, counteract global warming in the near term, and offset the Sun's gradual brightening over geological time-scales. Positioned at strategic points – such as the Sun–Earth Lagrange Point (L1) – these vast reflectors or thin-film shades would act like a planetary umbrella, shielding the planet without disrupting its ecosystems or weather systems.

1. Why use solar shading?

As the Sun brightens over millions of years, the solar flux (the amount of energy Earth receives) will eventually exceed the level that our climate systems can handle. Without intervention, this will trigger a cascade of environmental changes, ult-imately rendering our planet uninhabitable.

Reducing global temperatures

By blocking or reflecting a fraction of the sunlight before it reaches Earth's surface, solar shading could lower global temperatures and counteract the warming effects of the increased solar luminosity. This is critical, because unchecked heating will accelerate polar ice melt, raise sea levels, intensify extreme weather, and destabilize ecosystems.

Preventing runaway greenhouse effects

Rising carbon dioxide levels are already threatening to tip Earth toward a runaway greenhouse state – and increased solar radiation would amplify that risk. Solar shading could help maintain the planet's delicate energy balance, and prevent climate systems from spiraling out of control.

Extending Earth's habitability

This is not just about our survival; it is about taking control of our long-term destiny. With the right technology, we could keep Earth vibrant and habitable for billions of years. This would buy enough time for future generations to adapt, explore, and expand.

2. Methods for solar shading

Space-based sunshades

A giant sunshade at Lagrange Point L1 – about 930,000 miles (1.5 million km) from Earth – could intercept sunlight before it reaches us.

Options include:

- Thin reflective screens – vast, lightweight sheets to diffuse or reflect solar radiation.
- Swarms of small mirrors – billions of tiny satellites forming an adjustable, semi-transparent screen.
- Electromagnetic dust clouds – charged particles scattered into space to reflect sunlight.

The challenges are immense: materials will need to survive micrometeoroid impacts, intense radiation, and the vacuum of space, while maintaining precise orbital stability. Launching and assembling such a system would be one of the most expensive engineering projects in history.

Orbital mirrors and reflectors

Another option would be to place giant mirrors in orbit around Earth to deflect sunlight.

Designs might include:

- Satellites with adjustable reflectors – movable panels that redirect solar energy.

- Giant Mylar panels – expansive reflective surfaces deployed at key orbital points.

These would require precise navigation to maintain alignment and avoid interfering with astronomy, but they would offer flexibility and adaptability.

Stratospheric aerosol injection

Inspired by volcanic eruptions, this method would release reflective particles – such as sulfates – into Earth's upper atmosphere to scatter sunlight. This is a relatively low-cost solution, and it could provide short-term cooling.

But the risks include altering rainfall patterns and disrupting ecosystems. As with the other approaches under consideration, careful computer modeling and monitoring would be essential.

Cloud brightening

Spraying seawater into low-lying clouds could increase their albedo so they reflect more sunlight back into space.

This could cool down targeted regions, such as oceans or coastal cities, with fewer environmental risks than aerosols. However, its limited scalability makes it unsuitable as a sole global solution.

Manipulating Earth's albedo

Other approaches focus on increasing the reflectivity of Earth's surface. Examples include, painting rooftops white, installing reflectors in the deserts, or bioengineering plants to lighten their color. However, while these are reasonably cost-effective and easy to integrate with existing infrastructure, they only offer localized cooling and do not address the long-term challenge of solar brightening.

Protecting Earth from a supernova

A supernova is the explosive death of a massive star, or the catastrophic collapse of a white dwarf in a binary system. They are among the most violent and energetic events in the cosmos. In a matter of seconds, they can release energy equivalent to billions of suns, unleashing intense gamma-ray bursts and floods of high-energy particles.

If such an explosion occurred close to Earth, the radiation would strip away our ozone layer, leaving eco-systems exposed to lethal levels of ultraviolet light. The result could be mass extinctions and long-lasting ecological upheaval.

Earth's magnetic field, atmosphere, and ozone layer do a great job in protecting us from everyday cosmic radiation. But they would be powerless against the onslaught from a nearby supernova. Though such events are thankfully rare, we nevertheless need to develop planetary-scale defenses against this ultimate cosmic hazard.

As we saw in Chapter 1, IK Pegasi could erupt as a Type Ia supernova in the next 1.9 billion years, and it is close enough to Earth to cause global extinction.

Orbital radiation shields

One approach would be to build large-scale orbital shields to intercept or deflect radiation before it reaches Earth.

Possible designs include:

- Giant reflective screens – massive mirrors or panels stationed at the Lagrange points to scatter incoming radiation.

- Electromagnetic shields – vast superconducting structures that would generate artificial magnetospheres.

These would strengthen Earth's natural magnetic field and deflect high-energy particles.
We will study this option in more detail below.

- Dense particle shields – orbital clouds of dust or hydrogen-rich material that would absorb gamma rays before they hit the atmosphere.

These systems would require radiation-resistant materials, precise long-term orbital stability, and constant monitoring to counter gravitational drift. Building and maintaining them would demand unprecedented global cooperation and investment.

Stratospheric aerosol shields

A simpler – and more rapidly deployable – option would be to inject reflective aerosols into the stratosphere. This would scatter harmful radiation before it reached the surface.

This method is similar to the proposals for solar-shading against climate change, but it could help reduce the impact of a supernova's radiation.

However, aerosol shields also carry significant risks: they could disrupt rainfall patterns, alter water cycles, and cause ecosystems untold stress. They would also be far less effective against the most intense gamma-ray bursts.

This makes them a complementary or secondary measure rather than a standalone solution.

Underground and subsurface shelters

If a supernova threat were imminent, but we had sufficient notice, we could temporarily retreat into underground bunkers, subsurface cities, or even deep-ocean habitats. Rock and water provide excellent shielding against high-energy particles.

Backup settlements in lunar lava tubes or Martian caves could also extend protection beyond Earth. But building the infrastructure and moving entire populations underground (or into space) at short notice – probably a few decades – would be a colossal logistical challenge. Sustaining sealed ecosystems for long periods would also demand major advances in engineering and the management of resources.

Artificial magnetospheres

A more visionary defense involves the creation of artificial magnetospheres. These would consist of planetary-scale plasma shields that amplify Earth's natural magnetic field. We could use superconducting materials to generate powerful magnetic fields that deflect charged particles from the supernova.

These systems could also protect Earth against other cosmic hazards, including solar storms and cosmic rays, which makes them a versatile tool for planetary defense. However, the energy demands, materials science, and orbital deployment challenges are beyond our current capabilities.

Could we prevent a supernova?

In the far future, if we learn how to harness the full energy output of a star, we might be able to attempt stellar manipulation. We could then intervene before a star reaches the point of collapse.

Options could include:

- Gravity manipulation – using massive artificial objects to exert gravitational forces strong enough to alter a dying star's evolution.
- Energy transfer systems – siphoning off excess energy from a star to delay or prevent its collapse.

These ideas remain speculative, but if they were realized, they would give us the ability to defend ourselves and our planet against cosmic threats – and reshape the destiny of entire star systems.

★ Amazing future possibilities ★

Space-time topology computing: wormhole logic gates

Physicists have demonstrated, on small scales, the creation of wormhole-like structures within quantum computers. These structures allow for the teleportation of information through an abstracted extra dimension of space-time.

The experiments, inspired by the AdS/CFT correspondence and ER=EPR conjecture, suggest that space, time, and gravity might actually emerge from underlying quantum entanglement.

Projecting this forward, we can imagine that advanced civilizations might be able to construct space-time topology computers using artificial wormhole connections. They would then be able to manipulate not just information, but the very fabric of space-time as a computational resource.

Wormhole logic gates could, in theory, permit non-local computation, instantaneous data transmission across interstellar distances, or the simulation and manipulation of realities that are inaccessible in classical physics.

This could lead to models where information processing and fundamental physics merge, allowing computation not just *on* a substrate, but *of* the substrate itself.

Applications could include new approaches to dark matter and dark energy, revolutionary propulsion or teleportation systems, and hyper-secure communication – where intercepted data is literally in another space-time.

Astroengineering the Sun

If we were able to prevent a supernova from occurring, why stop there? In theory, that same mastery over stellar forces would enable us to prevent our own Sun from swelling into a red giant.

1. The Sun as a red giant

In approximately 5.4 billion years, the Sun will exhaust the hydrogen that powers its nuclear fusion. The core will collapse and heat up, igniting helium fusion.

> The helium will fuse into heavier elements including carbon and oxygen.

The intense heat will cause the Sun's outer layers to balloon outward, transforming it into a red giant. Its atmosphere will then extend beyond the orbits of Mercury and Venus, and could engulf Earth.

During this phase, the Sun will also lose mass through powerful stellar winds, creating a glowing planetary nebula. As a result, the Sun's gravity will reduce significantly, pushing the planets into wider orbits.

There is a slight advantage to this as it would give the planets a brief respite from the rising heat.

2. Preventing the red giant phase

Astroengineering the Sun to halt its evolution would require controlling the delicate balance between nuclear fusion and gravity – but on a stellar scale.

Removing mass from the Sun

A star's lifetime is determined by its mass: the more massive it is, the faster it burns through its fuel. Reducing the Sun's mass would prolong its hydrogen-burning phase and postpone the red giant stage.

Possible methods include:

- Solar mass-ejection devices – continuously extracting hydrogen from the outer layers.
- Controlled stellar winds – using engineered magnetic fields or plasma flows to accelerate the natural loss of mass.

If the Sun's mass were halved, it would bypass the red giant phase entirely, and evolve directly into a white dwarf.

However, removing mass on this scale would demand unprecedented technology and flawless precision. A miscalculation could destabilize the Sun or planetary orbits.

Artificially controlling fusion

Controlling the Sun's rate of fusion could extend its main-sequence life.

Potential approaches include:

- Injecting fresh hydrogen into the core to delay helium build-up.
- Suppressing fusion using exotic materials or magnetic confinement to slow down hydrogen consumption.

- Gravitational compression to stabilize the core without triggering helium fusion.

The precision required would be staggering – once again, even a small error could cause catastrophic instability.

Reconfiguring the Sun's internal structure

If we could intervene in the processes that form the helium core, we might be able to slow the expansion phase, or even halt it.

Ideas include:

- Restructuring plasma flows with massive electro-magnetic fields.
- Cooling the core to delay helium ignition and extend hydrogen fusion.

These interventions could, in theory, keep the Sun stable for tens of billions of years. However, its luminosity would continue to gradually rise, requiring additional climate-control measures on Earth.

3. Luminosity changes

Any modification to the Sun's mass or fusion processes would alter its brightness. This would have profound effects on planetary climates.

If mass were reduced

The Sun would burn through its hydrogen more slowly, gradually becoming cooler and dimmer. Its lifespan could stretch into the tens or hundreds of billions of years.

However, Earth would either need to move closer to the Sun to maintain its warmth, or rely on artificial climate systems.

If fusion were controlled

A steady, regulated fusion rate could keep the Sun's luminosity stable for billions of years, with a slow but predictable increase over time.

This would give us vast timescales in which to adapt – perhaps by gradually shifting Earth's orbit outward or by deploying solar-shading systems.

If hydrogen were replenished

Continuously refueling the core could keep the Sun in its main-sequence state indefinitely, potentially preserving stable conditions for trillions of years.

But harvesting, transporting, and injecting hydrogen on such as scale might be an even greater challenge than relocating the whole of humanity to another star system.

4. A visionary challenge in planetary protection

Preventing the Sun's red giant phase is one of the boldest ideas in astroengineering. While it could extend Earth's habitability for tens of billions of years, it would demand mastery of nuclear fusion, stellar mass, and plasma dynamics. We would also need breakthroughs in materials science and energy generation on an unimaginable scale.

For now, it remains a dream. But it's a dream that inspires us to think beyond the lifespan of our star, and to imagine a future where we not only survive the death of stars, but actually shape their destinies.

The Sun as a white dwarf

In around 7 billion years – unless we intervene – the Sun will complete its transformation into a white dwarf. This will be the final stage of its stellar life.

> A white dwarf is the dense, compact remnant left behind after a star sheds its outer layers and ceases nuclear fusion. The glowing core is composed mainly of carbon and oxygen.

It will be a beautiful and haunting sight, but one with profound consequences for the Solar System – especially for Earth and the Moon.

1. The collapse into a white dwarf

White dwarfs are extraordinary objects: roughly the size of Earth, yet containing the mass of the Sun. Their incredible density offers scientists a unique laboratory for studying the physics of matter under extreme conditions. This makes them invaluable to astrophysical research.

Once the Sun becomes a white dwarf, it will begin an immensely long cooling process. It will slowly fade over billions of years, gradually dimming as it radiates away the residual heat of its former life. Eventually, perhaps in trillions of years, it will become a black dwarf – a cold, dark stellar corpse that no longer emits light or heat.

Even then, its gravitational influence will persist, keeping the planets in their orbits. But those worlds will be frozen and lifeless, bathed only in the faint glow from other distant stars. But this final transformation lies so far in the future that it exceeds the current age of the universe.

2. The Sun's changing luminosity

The Sun's luminosity – its total energy output – is the key driver of Earth's climate and the Solar System's stability. Today, in its main-sequence phase, it radiates approximately 3.828×10^{26} watts.

During its red giant phase, that output will soar to 2,000–5,000 percent of its current level, as it swells and burns hotter.

But when it contracts into a white dwarf, its luminosity will plummet to roughly 0.0001 percent of today's value. It will shine faintly, emitting mostly infrared radiation, with only a weak trace of visible light. It will be a far cry from the bright yellow-white star we know today.

Implications for the Solar System

Such a dramatic drop in energy output would render Earth uninhabitable without extreme intervention. The oceans would freeze, ecosystems would collapse, and the biosphere as we know it would vanish.

To survive, we would have two primary options:

- Move Earth much closer to the white dwarf Sun, to a distance between 0.01 and 0.1 AU (roughly Mercury's current orbit or closer), so that it receives enough energy to sustain a stable climate.

- Rely on alternative energy sources, such as large-scale nuclear fusion or Dyson-sphere collectors around other stars, to replace the lost solar input.

If Earth were moved inward, the consequences would be profound. For example, a year would shrink to just 88 days (or fewer), and we would need to adapt not only our calendars but our agriculture and our entire way of life.

Re-engineering Earth

Imagine a future in which Earth becomes a fully engineered world. It would be a deliberate creation of human ingenuity, designed to sustain life indefinitely. Every fundamental system – climate, atmosphere, energy, ecosystems, and even the planet's geological processes – would be reshaped to form a permanently stable, self-sustaining biosphere. It would flourish regardless of any cosmic threats, including the eventual fading of the Sun.

1. Transforming Earth's atmosphere and climate

As the Sun brightens, and later dims into a white dwarf, Earth's natural systems will face extreme shifts that threaten its habitability. To ensure that life continues, we would need to take full control of the planet's climate and atmosphere, replacing the Sun's influence and directly regulating the weather systems.

Replacing the Sun's influence

The Sun's role as our primary energy source could be replicated using planetary-scale technologies that provide heat and light.

- **Artificial lighting grid**

  A vast orbital network of solar mirrors or artificial light sources could deliver consistent illumination worldwide.

  Adjustable reflectors would preserve the natural day–night cycle, balance seasonal differences, and direct light where it is most needed.

- **Fusion-powered atmospheric heating**

  Fusion reactors – which mimic the Sun's energy generation process – could emit infrared radiation to warm the surface and atmosphere evenly.

  Integrated thermal-management systems would enable precise control of the climate.

Artificial weather control

Without the Sun's natural drivers, we would need to manage Earth's atmospheric processes directly.

- **Terraforming-based climate regulation**

  Advanced weather machines could generate clouds to cool overheated regions, trigger controlled rainfall to maintain water supplies, and suppress storms and destructive weather events by stabilizing pressure systems.

- **Magnetic field strengthening**

  Artificial electromagnetic shields could reinforce the magnetosphere, protecting the climate from cosmic radiation and charged particles as the Sun's magnetic influence wanes.

2. Engineering Earth's ecosystem

A stable, resilient biosphere will form the foundation of our long-term survival. By combining biotechnology, robotics, and climate management, we could preserve biodiversity indefinitely, and create adaptive, self-sustaining environments.

Bioengineered plants and animals

Genetic engineering could produce species that can:

- thrive in low-light conditions
- tolerate altered atmospheric compositions
- withstand higher or lower levels of carbon dioxide

Enhanced photosynthesis would enable crops to produce food and oxygen under diminished sunlight. Bioengineered microbes could restore soil fertility, regulate oxygen, and process waste.

Cybernetic wildlife

Hybrid biological–robotic species could perform essential ecological roles:

- cybernetic bees for pollination
- semi-artificial fish to clean the oceans
- adaptive organisms to monitor and maintain environmental balance, and act as custodians of the biosphere

Closed-loop biosphere systems

- Artificial forests with high-efficiency photosynthetic structures could act as oxygen factories, producing breathable air and absorbing carbon dioxide.
- Oceanic maintenance systems could regulate acidity, filter out pollutants, and circulate nutrients to sustain marine life.
- Bioengineered coral reefs could resist warming and acidification, preserving marine biodiversity.

3. Geological stability and terraforming

Over geological timescales, Earth's tectonic activity will slow down, weakening its ability to regulate carbon cycling and generate a magnetic field.

Artificial intervention could maintain geological stability and habitability.

Artificial core heating

Geothermal reactors installed deep in the crust could transfer heat to the core, sustaining the convection currents and the geodynamo that generates Earth's magnetic field. This would also stabilize tectonic processes and provide abundant geothermal energy.

Reinforced geological stability

- **Active earthquake dampening**

  Subterranean shock absorbers, vibration-damping layers, and AI-controlled seismic systems would prevent destructive quakes.

- **Artificial plate tectonics**

  Controlling mantle convection or creating new subduction zones would maintain carbon recycling and atmospheric stability. Again, this would need to be carefully controlled to prevent earthquakes.

Interventions like these would preserve Earth's ability to self-regulate its climate, and shield life from cosmic hazards. Together, they should ensure planetary stability for billions of years.

4. Energy sources for an engineered Earth

Replacing the Sun as Earth's primary source of energy would be one of the most ambitious undertakings in our history – including our future yet to come. But we would have the advantage of billions of years to prepare.

The solutions we might develop – from planetary-scale fusion grids to cosmic-scale energy harvesting – will demonstrate the ingenuity and the audacity of human innovation.

Fusion energy grid

A network of planetary-scale fusion reactors could replicate the Sun's nuclear processes on Earth. They would provide a stable, virtually limitless supply of clean energy. Fueled by isotopes such as deuterium and tritium – which are abundant in the oceans – these reactors would sustain controlled fusion to generate immense amounts of heat and electricity.

Thousands of interconnected fusion plants would form a distributed grid, ensuring that energy reaches every region – with backup systems in case of maintenance or disruption. Output could be adjusted dynamically to match the needs of different climates, industries, and ecosystems, so that all of Earth's engineered systems continue to run smoothly.

Dyson-like orbital power stations

While fusion could form the backbone of the planetary energy grid, Dyson-like orbital stations could complement it by capturing energy on a cosmic scale.

Massive solar collector satellites stationed around nearby stars could beam their harvested light back to Earth via microwaves or lasers.

Some might have giant mirrors to redirect the starlight to Earth, compensating for the loss of our Sun. Equipped with precision aiming systems, these reflectors could focus light and warmth on specific regions. Combined with Earth-based systems, they would form a resilient, multi-layered energy infrastructure capable of sustaining the planet indefinitely.

Exotic energy sources

Looking further ahead, we might one day be able to tap into energy sources that remain purely theoretical today.

For example, dark matter reactors might exploit the properties of the universe's most elusive substance to provide practically infinite power. Advances in quantum energy manipulation might allow us to draw energy from the quantum vacuum, or harness other untapped physical phenomena. This would entirely redefine our relationship with energy.

5. Social and civilizational adaptation

Transforming Earth into a fully engineered world, independent of the Sun's natural energy, would demand not only technological breakthroughs but also profound cultural and societal shifts. Thriving in such an environment would require us to evolve in how we live and govern. It could even redefine what it means to be human.

Post-biological evolution

Future human generations might transition into post-biological forms, merging biology with advanced technology. Digital consciousness would allow individuals to upload their minds into virtual environments, where they could live without the constraints of biology or physics. Others might adopt bio-mechanical forms, becoming cybernetic hybrids with enhanced durability and adaptability. They would be able to survive in low light, high radiation, or extreme or altered atmospheres.

Self-sustaining megacities

As ecosystems become increasingly artificial, vast enclosed megacities could provide us with stable, independent habitats. These enormous cities would recycle air, water, and waste; generate their own renewable energy; and simulate the natural rhythms of day and night. Artificial skies would display dynamic clouds, sunlight and stars, preserving psychological well-being in a world without a natural Sun.

These vibrant cities could host millions of people, while remaining entirely autonomous from the outside environment.

Ultra-long-term governance

Managing an engineered Earth over astronomical timescales would require governance systems capable of enduring for millennia. Advanced AI could oversee planetary operations, monitoring climate control, resource distribution, and population dynamics. Such a system would need to be precise and impartial, and handle vast quantities of data.

Interplanetary trade networks could link Earth with colonies across the Solar System and beyond, exchanging resources, knowledge, and technology. These connections would advance our cultural and technological progress, and help unite humanity as a single, distributed civilization.

If these breakthroughs were achieved, Earth could remain a thriving home indefinitely, even with a geologically dormant core and in a universe without sunlight.

An engineered Earth would be a true testament to our imagination, our determination, and our will to endure.

Ecosystem self-repair systems

In a fully engineered Earth, where natural processes no longer operate independently, self-repair systems would be essential. These advanced systems would maintain planetary stability and long-term sustainability, replicating nature's regenerative capabilities.

Together, they would enable biospheres, oceans, forests, and atmospheric cycles to maintain themselves.

1. Automated carbon dioxide and oxygen regulation

Maintaining a breathable atmosphere means keeping the delicate balance between oxygen and carbon dioxide in check. Without natural photosynthesis to perform this role, we would need automated systems to take over.

Synthetic photosynthesis and carbon capture

Artificial trees and algae towers could draw carbon dioxide from the air and convert it into oxygen, purifying the atmosphere while reducing greenhouse gas levels. We could deploy these system in any region that needed them, from cities to deserts, making them highly scalable.

Nanotechnology could produce catalysts that absorb carbon dioxide directly from the air and break it down into harmless components, operating continuously to prevent dangerous build-ups. In the oceans and soil, bioengineered microbes could metabolize excess carbon dioxide to improve water and soil quality while supporting the global carbon cycle.

Self-maintaining oxygen generation

Genetically engineered plants could be developed to thrive in extreme conditions – low light, minimal water, or high carbon dioxide – while producing abundant food and oxygen.

In deep-sea habitats or subterranean colonies, oxygen-producing bioreactors could simulate photosynthesis using engineered microorganisms or algae.

Within sealed megacities, automated air-filtration systems would monitor and adjust oxygen and carbon dioxide levels in real time. They would adapt to population changes within moments, ensuring optimal air quality for the millions of residents.

2. Soil and agriculture self-restoration

To feed the global population and sustain the planet's ecosystems, we will require soils that can renew themselves indefinitely.

Bioengineered microorganisms for soil health

Self-renewing bacteria and fungi could act as living fertilizers. They would break down organic matter, fix nitrogen, and recycle minerals without the need for chemical additives. Smart microbes could detect pollution, erosion, or nutrient deficiencies and release targeted compounds to restore the balance, working autonomously to keep farmland fertile.

Automated nutrient cycling

Nanobot networks could monitor soil chemistry and deliver minerals precisely where they are needed. AI-driven farming systems could adjust irrigation, nutrient delivery, and pest control dynamically, optimizing yields with minimal human intervention.

3. Ocean ecosystem self-regulation

The oceans are central to Earth's climate and life cycles. They act as carbon sinks, temperature regulators, and biodiversity hotspots.

Self-replenishing coral reefs

Genetically modified coral, resistant to heat and acidification, could regenerate rapidly and restore damaged reefs. These resilient reefs would revitalize marine habitats and support the countless species that depend on them.

Other bioengineered marine life – including disease-resistant fish, shellfish, and plankton – could stabilize food chains and nutrient cycles, strengthening the web of life in the seas.

Ocean cleanup and pollution absorption

AI-equipped marine drones could patrol the oceans, filtering out plastics, toxins, and debris on a massive scale.

Algae-based scrubbers could absorb excess carbon dioxide from seawater, reducing acidification and protecting sensitive marine species.

Together, these systems would restore water quality and safeguard our oceans' health.

4. Climate stability and weather control

Smart climate regulators

Atmospheric drones could monitor and adjust cloud formations to control rainfall – for example, triggering showers in drought zones or dispersing clouds to prevent floods. Ozone-repair systems could deploy engineered molecules to repair any damage in the vital ozone layer.

Natural disaster prevention

We have looked at ways of preventing damaging storms. And we have seen how seismic stabilizers embedded deep in the crust could monitor tectonic activity and apply counter-forces to prevent major earthquakes or tsunamis.

Coastal tsunami barriers could deploy automatically, redirecting dangerous waves away from populated areas.

These tools would give us unprecedented control over Earth's most volatile forces, ensuring a stable, habitable planet for the long term.

Perhaps the only things we might not be able to fully control are volcanoes. But we might be able to tap into the immense reservoirs of energy beneath them, and this might lessen their impact when – or if – they erupt. It would also provide us with a valuable source of heating and power.

5. Biodiversity and artificial species management

Earth's natural ecosystems are under unprecedented pressure from environmental change, habitat loss, and accelerating extinction rates. Engineered ecosystems offer a powerful way of preserving biodiversity and preventing species' decline. By combining cutting-edge biotechnology with advanced artificial intelligence, we could create dynamic, resilient environments, and ensure that life on Earth continues to flourish for millennia to come.

Synthetic wildlife and bioengineered species

Synthetic and bioengineered species could transform how we restore and stabilize ecosystems. For example, AI-monitored wildlife zones could track populations, migration patterns, and ecological health in real time. These intelligent systems would respond instantly to imbalances. They could introduce new species, adjust resources, or manage habitats – without the need for human intervention.

Artificial, self-replicating plants could help reclaim devastated landscapes. They would anchor the soil, generate oxygen, and provide food and shelter for other species. Designed to thrive in extreme environments – from arid deserts to post-industrial wastelands – these plants could jump-start the recovery of entire ecosystems.

Together, these innovations would allow us to preserve not just individual species, but the intricate web of interconnections that sustains life.

Genetic vaults for long-term survival

Genetic vaults could act as time capsules for life, storing seeds, DNA, and even living cells from countless species. These repositories would safeguard the genetic blueprints of Earth's biodiversity, and could be used to revive lost species or rebuild damaged ecosystems after an environmental catastrophe.

Facilities like the Svalbard Global Seed Vault in Norway already demonstrate the feasibility of this approach.

In the future, we could take this concept further with automated evolution simulations. These predictive systems would model how species might adapt to changing conditions, such as rising temperatures, shifting habitats, or new ecological pressures.

By pre-adapting species before they are reintroduced, and optimizing them for survival in transformed environments, we would ensure they continue to thrive – no matter what challenges a fully engineered Earth presents.

Speculative astrophysics and the future of human survival

As we push the boundaries of scientific discovery, mastering space-time manipulation, dark energy, and exotic physics could become essential. They will enable interstellar travel, survival beyond Earth, and vast engineering projects on a cosmic scale.

While these concepts remain theoretical today, future breakthroughs in physics could transform them into practical realities.

1. Exploiting dark energy

Dark energy – the mysterious force that drives the accelerating expansion of the universe – is thought to make up around 68 percent of the cosmos's total energy content. If we could find a way of harnessing it, the implications for energy sustainability, space travel, and our civilization's future would be staggering.

Vacuum energy extraction

Dark energy might originate from quantum fluctuations in the vacuum of space. This omnipresent energy field is woven into the very fabric of the universe. If we could tap into it, we could generate unlimited power. We could also build spacecraft with continuous-thrust drives that never run out of fuel. This would enable deep-space colonization without the logistical constraints of conventional propellants.

Negative energy applications

Negative energy – the theoretical counterpart to positive energy – could stabilize warp fields and wormholes, making faster-than-light travel feasible. It might also counteract gravitational collapse, allowing long-term stabilization of space-time distortions.

Mastering gravitational control could enable planetary-scale engineering, including:

- Building habitats in environments once deemed uninhabitable.
- Creating self-sustaining ecosystems on rogue planets that drift through space without starlight.
- Establishing artificial gravitational constructs in deep-space voids, near neutron stars, or in entirely engineered habitats.

Such amazing capabilities would free us from the narrow confines of traditional habitable zones.

2. Exploring the multiverse

The idea of parallel universes – alternate realities where different versions of existence unfold – has long fascinated science, philosophy and literature. While parallel universes are currently theoretical and unproven, quantum mechanics offers us tantalizing hints that they might exist.

If future civilizations could cross between them, the implications for our survival and technological evolution would be beyond anything we can currently imagine.

Quantum tunneling into parallel universes

Quantum theory suggests that alternate realities could arise from cosmic inflation, branching timelines, or higher-dimensional structures.

If we could quantum-tunnel between them, inter-dimensional travel might become possible. We could use this as a survival strategy – an escape route from cosmic catastrophes such as vacuum decay or even the heat death of our universe.

Multiversal travel could also allow us to explore diverging versions of Earth. For example, we could witness alternate evolutionary paths, societal developments, and scientific possibilities that never unfolded in our own timeline.

Transferring consciousness into an energy-based existence

Another possibility is the migration of human awareness into a non-material, energy-based form, such as light, plasma, or quantum information. Freed from biology and traditional physics, we could exist indefinitely, immune to aging, injury, and disease.

Intelligence in this state might function as pure thought, interfacing directly with quantum fields and traversing space as encoded bursts of information.

Such beings might communicate instantaneously across universes, experience time non-linearly, and be able to perceive multiple realities at once. Civilization might expand across multiversal dimensions, giving us a god-like omnipresence.

But this raises profound philosophical questions:

- Does identity persist when it is detached from its physical form?
- Would individuality dissolve into a collective hive-like mind, or could autonomous thought survive in a quantum-based existence?

We may not be able to answer these questions for millions of years. But one day they could define the ultimate trajectory of human evolution.

Innovation has carried us far, but our true adventure will begin when we leave the Sun behind. Beyond the Solar System lies the unknown: starships, new forms of humanity, civilizations beyond biology, and the possibility of neighbors among the stars.

Part III

Beyond the Solar System

The night sky becomes a map; the map becomes a plan.

Engines light in the dark. Generations commit to journeys that only their descendants will complete.

Leaving the Sun's warmth and light behind, we learn what in us endures.

7

Interstellar travel

Generational ships, cryogenic stasis, and warp drives could potentially carry us and our descendants to distant stars. But each method has its own technical, social, and ethical challenges.

The staggering distances between solar systems

Discovering habitable exoplanets is a thrilling scientific milestone, but reaching them is an entirely different challenge. The distances between solar systems are so vast that they defy everyday comprehension. Reaching them will push our ingenuity to its absolute limits.

Even within our local cosmic neighborhood, the journey to other stars will require breakthroughs far beyond today's propulsion technology. As we saw earlier, researchers hope to launch a fleet of nanoprobes toward Alpha Centauri and Proxima Centauri in around 2036, with the aim of arriving in 20 to 30 years. But with our current technology, a larger, crewed spacecraft would take tens of thousands of years to make the same journey.

This immense scale will force us to rethink space travel. Our desire to reach these far-flung targets will drive the development of advanced propulsion systems, the discovery of new kinds of physics, and the execution of engineering projects on a scale we have never attempted before.

1. Measuring interstellar distances

The enormous vastness of space is almost impossible to visualize. Astronomers measure the distance between stars in light-years. This is the distance light travels in one year – which is roughly 5.88 trillion miles (9.46 trillion kilometers).

Proxima Centauri, our nearest stellar neighbor, is 4.24 light-years away – or about 24.93 trillion miles.

To put this into perspective, the farthest human-made object, *Voyager 1*, launched in 1977, is traveling at 38,000 mph (10.5 miles per second) and is now around 16 billion miles from Earth. In light-year terms, that is essentially zero – or, more precisely, about 0.00272 light-years.

The signals it sends back to Earth take around 24 hours to arrive, meaning that it's about one light-day away. So, on its roughly 50-year journey, it has traveled about 1/365th of a light-year – or 1/1,548th the distance to Proxima Centauri.

Our nearest stars

The three closest stars – Proxima Centauri, Alpha Centauri, and Barnard's Star – lie between 4 and 6 light-years from Earth. At *Voyager 1*'s current speed, the journey to Proxima Centauri would take over 73,000 years. So even stars that are nearby in cosmic terms are unimaginably far away in human terms.

The expansive Milky Way

Our galaxy spans more than 100,000 light-years. *Voyager 1* would take nearly 2 billion years to cross it.

Our earliest radio broadcasts, sent just over a century ago, have traveled only around 1/1,000th of the way across the Milky Way. This means that 99.9 percent of any potential civilizations in our galaxy probably have no idea we exist – our broadcasts have simply not reached them yet.

Another issue is that those early transmissions are now so faint that they are virtually indistinguishable from the cosmic background noise. Only our most powerful signals – such as the three-minute "Arecibo message" we broadcast into space in 1974 – will remain detectable beyond a few hundred light-years. Airport radar might be detectable within a range of about 200 light-years, but since those systems only began operating in the 1950s, their signals have only traveled around 75 light-years so far.

> If a civilization 75 light-years away detects one of these signals and decides to send a response, it will take another 75 years for their message to reach us.

Beyond the galaxy

The nearest major galaxy, Andromeda, lies about 2.5 million light-years away. This means that the light we see shining from it today began its journey to Earth around 2.5 million years ago – long before modern humans evolved. This is another humbling reminder of the vast scale we are dealing with.

2. The challenges of crossing interstellar space

If we are to make interstellar travel a reality, we will need to overcome fundamental barriers in propulsion, radiation protection, and energy sustainability.

The limitations of current space travel

Today's rockets rely on chemical propulsion. This is powerful enough for missions within our own Solar System, but hopelessly inefficient across interstellar distances.

Chemical rockets require enormous amounts of fuel, and are far too slow to make journeys between the stars.

An interstellar spacecraft would also need to survive for several decades – or perhaps even centuries – in deep space. Outside of Earth's magnetic shield, travelers would face intense cosmic radiation and high-energy particles from distant stars. Energy supplies would have to last for the entire voyage, with no chance of resupply.

The future of interstellar travel

To succeed, we will need to develop propulsion systems that break free from these constraints. Potential concepts include light sails, fusion drives, antimatter engines, and possibly even warp-drives. Each breakthrough we make will take us one step closer to turning interstellar travel from science fiction into reality.

The challenge is immense, but so too is our drive to explore. The real question is not *whether* we will cross the immense gulf between the stars, but *when* we will achieve it.

How to reach Proxima Centauri b

Astronomers have seen hints that Proxima Centauri b could be habitable, but with today's technology we are unable to confirm it. To know for sure, we would need to visit it, take close-up images, and gather direct measurements. Such a feat will remain out of our reach until we develop propulsion systems far beyond anything we have now.

Let's take a look at the best propulsion systems available today, and the revolutionary concepts that might one day power us – or, more likely, our distant descendants – to Proxima Centauri b and beyond.

Ion propulsion

Ion propulsion is one of the most efficient propulsion methods currently in use. It trades raw power for extraordinary endurance. Unlike chemical rockets, which burn fuel in short, powerful bursts, ion engines generate a weak but continuous thrust by accelerating charged particles to extreme speeds.

How it works

- Xenon gas, chosen for its inert nature and ease of ionization, is fed into the thruster.
- Electrons bombard the xenon atoms, stripping away their electrons to create positively charged ions.
- These ions are accelerated through an electrically charged grid and expelled at speeds of around 67,000 mph (108,000 km/h).
- Thrust is produced through Newton's Third Law: every action has an equal and opposite reaction.

The thrust is gentle but relentless, which makes ion propulsion ideal for long-duration missions where efficiency matters more than raw acceleration.

Real-world use

NASA has already proven the concept:

- *Deep Space 1* (1998) tested ion propulsion while exploring asteroids and comets.
- *Dawn* (2007) used it to visit Vesta and Ceres in the asteroid belt – a feat that chemical rockets could not have achieved in a single mission.

The biggest drawback is that – at their current levels of performance – ion-powered craft would take 81,000 years to reach Proxima Centauri b.

Clearly, we need something much faster.

Gravity assist

Also known as the slingshot maneuver, gravity assist uses the motion and gravity of planets or moons to boost a spacecraft's speed and alter its trajectory. It can achieve this without burning any extra fuel.

How it works

- The spacecraft approaches a planet or moon, and falls into its gravitational field.
- As it swings past, it accelerates, stealing a tiny fraction of the celestial body's orbital momentum.
- It departs at a higher speed – and often on a new trajectory.

Real-world use

- *Voyager 1* and *Voyager 2* (1977) used Jupiter and Saturn to escape the Solar System.
- *Galileo* (1989) used Venus and Earth to reach Jupiter.
- *Cassini-Huygens* (1997) used Venus, Earth, and Jupiter to reach Saturn.
- *New Horizons* (2006) shaved years off its trip to Pluto by swinging past Jupiter.
- The *Parker Solar Probe* (2018) used Venus – not to speed up, but to slow down enough to approach the Sun.

A carefully planned series of gravity assists could reduce the travel time to Proxima Centauri b from 81,000 years to around 19,000 years. While this is a massive reduction, it is still far too long for human exploration.

The need for revolutionary propulsion

Our aim is to reach Proxima Centauri b within a single human lifetime, and for that we will need propulsion systems that break free from current limitations.

Light sails

Also called solar sails, these spacecraft harness the pressure of light itself. While photons have no mass, they do carry momentum, and when they strike a reflective surface, they impart a tiny push. Over time – and billions upon billions of photons – that push can build into astonishing speeds.

How it works

- A light sail is made from an ultra-thin, highly reflective material such as Mylar or aluminum-coated plastic.
- The sail captures photons from sunlight or from powerful Earth-based lasers.
- In the vacuum of space, with no drag to slow it down, this faint but constant push can accelerate a craft to a significant fraction of light speed.

Because light sails carry no on board fuel, they are incredibly lightweight and could, in theory, continue to accelerate for many years.

Pioneering missions

- *IKAROS* (Japan, 2010) became the first deep-space solar sail mission, successfully propelling itself toward Venus.
- *LightSail 2* (The Planetary Society, 2019) demonstrated controlled solar sailing in Earth orbit.
- *Breakthrough Starshot* (planned for around 2036) aims to use high-powered lasers to propel a fleet of 1,000 tiny, sail-equipped nanocraft toward Alpha Centauri and Proxima Centauri, potentially arriving in just 20 to 30 years.

Challenges and potential

Acceleration is gradual, so this method demands immense patience. Sail materials will need to survive decades of radiation and micrometeoroid impacts. But, if we can perfect this, light sails could form our first practical step toward interstellar travel. Indeed, it could be the technology that carries the first human-made spacecraft to another star.

Fusion propulsion

Harnessing the power of controlled nuclear fusion could transform space travel, opening the door to speeds of around 10 percent of light speed.

While this is only half the theoretical top speed of light sails, fusion technology would be able to propel much larger, heavier spacecraft – including crewed missions. At this speed, we could reach Proxima Centauri in 40 to 50 years, achieving our aim of interstellar travel within a single human lifetime.

How it works

Fusion propulsion is based on the same process that powers the stars. Hydrogen isotopes – typically deuterium and tritium – are forced together under extreme heat and pressure. When their nuclei fuse, a small portion of their mass is converted directly into energy – as described by Einstein's famous equation, $E = mc^2$.

This reaction produces high-energy particles, such as helium nuclei and neutrons, which can be channeled into thrust.

Scientists are studying two main approaches:

- Magnetic confinement propulsion, which uses magnetic fields to direct charged fusion particles and generate thrust.
- Plasma expansion propulsion, which uses fusion energy to heat plasma, which then expands and is expelled at high velocity.

Why fusion outperforms chemical rockets

Fusion produces vastly more energy per unit of fuel than chemical combustion, and it is far more efficient. Deuterium and tritium are relatively abundant and compact, meaning that spacecraft could carry enough fuel for decades-long missions without the colossal mass of chemical propellant.

The challenges ahead

While the promise is huge, so too are the hurdles. Fusion requires temperatures of tens of millions of degrees Fahrenheit to overcome the natural repulsion between atomic nuclei. This is far beyond the temperature that current engineering can sustain in a stable, compact reactor.

Other obstacles include:

- Efficiently converting fusion energy into usable thrust.
- Building reactors that can operate reliably in the harsh environment of space.
- Protecting spacecraft from the damage caused by high-energy neutrons.

Proposed concepts

- Princeton Satellite Systems' Direct Fusion Drive is a compact system that uses magnetic confinement to produce both thrust and electrical power.
- Project Daedalus (British Interplanetary Society) is a pulsed-fusion starship concept designed to reach nearby stars within a single human lifetime.
- Helion Energy's Helion Fusion Engine is an experimental reactor that could be adapted for propulsion.

The future

If we can overcome the challenges, fusion propulsion could become the workhorse of interstellar travel. It is efficient, long-range, and powerful enough to carry us to other stars.

Antimatter propulsion

Imagine a spacecraft accelerating toward the stars at close to light speed, reaching Proxima Centauri b in just five years, and any star within 50 to 60 light-years of Earth in a single human lifetime.

This is the promise of antimatter propulsion – a theoretical technology that uses the most efficient energy source known: the annihilation of matter and anti-matter.

The physics

When matter meets antimatter, they annihilate completely, converting 100 percent of their mass into energy. This is far more efficient than fusion, which converts only a fraction.

Antimatter particles are the mirror opposites of normal matter: positrons are the antimatter equivalent of electrons, and antiprotons are the antimatter equivalent of protons. When they collide, they release intense gamma rays and high-energy particles.

How it could work

The two main proposals mirror those of fusion propulsion:

- Magnetic confinement propulsion directs charged annihilation products through magnetic fields to produce thrust.
- Plasma expansion propulsion uses annihilation energy to heat plasma, which expands and is then expelled at extreme velocities.

As the energy yield is so high, only small amounts of antimatter would be needed for interstellar missions. This would make these journeys dramatically faster and more practical than with fusion.

The challenges

While the science is solid, the engineering challenges are formidable.

- **Production**
  Current facilities like CERN can only produce minute quantities of antimatter – nowhere near enough to propel a spacecraft. It currently costs around $62.5 trillion to produce a single gram, and a journey to Proxima Centauri and back within a single human lifetime would likely require thousands of tons.

- **Storage**
  Antimatter cannot touch normal matter without annihilating it, so it must be suspended in magnetic traps inside vacuum chambers.

- **Containment risks**
  Even a microscopic leak would cause a catastrophic explosion.

- **Energy conversion**
  Converting annihilation energy into controlled thrust requires technologies we have not yet developed.

Research and concepts

NASA has explored the theory of antimatter propulsion, including the use of antiprotons to trigger nuclear reactions.

Project Valkyrie is another concept: a crewed starship powered by antimatter engines that could potentially reach 92 percent of light speed.

Outlook

While we are still a long way from building a working antimatter engine, the physics behind it is one of the most exciting frontiers in propulsion research. If we can solve the problems of production, containment, and energy conversion, we could achieve interstellar travel to several hundred stars within a human lifetime.

Warp drives

Warp drives are among the boldest ideas in theoretical physics. They will enable us to cross vast interstellar distances in what would feel like moments – without breaking the fundamental limit of light speed.

The concept does not require a spacecraft to move faster than light in the traditional sense. Instead, it manipulates space-time, allowing the ship to "ride" a distortion in the fabric of the universe and arrive at a distant star in a fraction of the time conventional travel would take.

How it could work

A warp drive would create a "warp bubble" around the spacecraft. Inside the bubble, the ship would remain stationary relative to its immediate surroundings. The bubble compresses space-time ahead of the craft and expands it behind, effectively carrying the ship forward as space itself moves.

As the ship is not accelerating in the usual way, passengers would not experience the extreme time dilation that comes with near-light-speed travel. In fact, time on board would pass normally.

The role of exotic matter and negative energy

To generate a warp bubble, the drive would require exotic matter or negative energy – substances that behave in ways that are unknown in ordinary physics. In theory, these could bend space-time in the precise way needed. Some physicists suggest that quantum vacuum fluctuations might produce the necessary negative energy, but no practical method of harnessing it has yet been found.

The Alcubierre model

In 1994, Mexican theoretical physicist Miguel Alcubierre proposed a mathematical framework for such a drive, based on Einstein's theory of general relativity. His equations showed that space-time distortions could, in theory, allow for faster-than-light travel. The catch is that the amount of energy required for a single journey would be staggering – perhaps more than the total mass-energy of an entire planet.

Challenges and risks

Even if we could solve the energy problem, warp drives would face several other daunting hurdles:

- Generating and maintaining a stable warp bubble is purely hypothetical.
- If the bubble collapsed, it could destroy the spacecraft.
- Interactions between the bubble and surrounding space-time are unpredictable and could affect nearby objects.
- Exiting a warp bubble might release dangerous bursts of radiation.

NASA's Eagleworks Laboratory has conducted initial theoretical studies into space-time manipulation, but for now, warp drives remain a distant prospect.

Quantum mechanics

Quantum mechanics describes the strange, counter-intuitive behavior of matter and energy at the smallest scales. It underpins much of modern technology, and might one day lead to propulsion methods that change how we explore the galaxy.

Fundamental phenomena

At the quantum level, particles can behave in ways that defy classical expectations. In theory, we could harness some of these effects for space travel.

- **Quantum entanglement**
  When two particles become linked, the state of one instantly influences the other, regardless of the distance between them. This could allow for instantaneous communication across interstellar distances – which will be a vital technology for deep-space missions.

- **Quantum tunneling**
  Particles can sometimes pass through barriers they should not be able to cross. This effect, seen in nuclear fusion and electron behavior, might one day lead to propulsion systems that "shortcut" through space-time.

- **The Casimir effect**
  Two closely spaced plates in a vacuum can generate negative energy, which is a theoretical requirement for stabilizing warp bubbles and wormholes. If this negative energy could be harnessed, it might allow us to manipulate space-time.

While these ideas are highly theoretical, quantum mechanics continues to challenge our understanding of what is – or might be – possible.

Wormholes

Wormholes (also called Einstein-Rosen bridges) are hypothetical tunnels through space-time that connect two distant points. They could allow instantaneous travel between those points, without the need to cross the space between them.

Some theories suggest they could also link different points in time, or even parallel universes. This would open up some amazing possibilities for future exploration.

How they might work

A wormhole would have two "mouths" connected by a short passage through space-time. In some models, this passage (also called a tunnel or throat) is so short that travel between the mouths would effectively be instantaneous – even if the mouths were billions of light-years apart.

The challenges

- **Exotic matter and negative energy**
  We would need these exotic phenomena to keep the wormhole stable, but they have never been observed in nature. Without them, the wormhole could collapse before it could be used.

- **Scale**
  Theoretical wormholes might only exist at subatomic sizes. Enlarging one to fit a spacecraft would require technology far beyond our current capabilities.

- **Extreme forces**
  The gravitational and radiation effects near a wormhole's throat could be lethal without levels of shielding that are also far beyond our current technology.

The paradox of time travel

If wormholes allow for travel between different points in time, this could raise deep logical problems. This includes the famous "grandfather paradox," where altering the past could erase the conditions that allowed the time travel event to happen in the first place.

There are three popular theories on the effects of traveling into the past.

1. The Grandfather Paradox suggests that if you traveled into the past and killed your grandfather, then you would not have been born. Therefore, you could not have traveled into the past and killed your grandfather.

2. The theory of no effective change suggests that events will always occur as they are recorded in history. For example, if you traveled into the past and killed Adolf Hitler when he was a child, then his parents might have adopted a child who subsequently grew up to become Adolf Hitler – or committed the same atrocities.

3. The theory of branching timelines suggests that when you travel into the past and cause a change, you create a new timeline. Events will continue along the original timeline as if nothing had happened. Events will be different on the new timeline, which runs in parallel with the original one. But you can never jump from the new timeline to the original one.

The future of wormhole research

Physicists are exploring whether quantum principles and space-time manipulation could stabilize wormholes.

NASA's Eagleworks team, which has studied warp drive concepts, has also examined wormhole theory. Advances in quantum computing and entanglement could one day help us understand space-time geometry well enough to test these ideas for real.

What if we cannot go that fast?

If near-light-speed or faster-than-light propulsion remain out of our reach, we still have other options. There are concepts rooted in known physics that could take us to the stars – though they would demand patience on a scale few of us can imagine.

Generational ships

Also called world ships or interstellar arks, these are among the most ambitious visions for deep-space travel. Instead of racing to another star in just decades, they would carry human civilization on voyages that last centuries, or perhaps even millennia. These would be entire, self-contained societies, preserving knowledge, culture, and the human spirit as they travel between the stars.

A new kind of civilization among the stars

Unlike conventional missions, where crews depart for space and return to Earth within their own lifetimes, a generational ship would be home to communities who live and die without ever seeing the world their ancestors left – or the one their descendants will colonize. Entire generations would be born, educated, work, reproduce, and grow old within the vast, artificial environment of their vessel.

Sustaining life for centuries

To succeed, such a ship would need to operate as a closed-loop ecosystem, recycling air, water, and nutrients almost indefinitely.

- **Food**
  Advanced farming methods such as hydroponics, aquaponics and aeroponics would feed the population.

- **Waste**
  Organic material would be converted into compost or bioenergy.

- **Protection**
  Heavy shielding would guard against cosmic radiation, micrometeoroids, and other hazards.

- **Power**
  Long-duration energy sources such as nuclear reactors would be essential. They could be supplemented by solar sails or solar panels when the ship is near a star.

- **Gravity**
  Most designs feature rotating habitats to create artificial gravity via centrifugal force. This would prevent the muscle and bone loss that would otherwise plague long-term spacefarers.

> **Hydroponics** is a method of growing plants in water laced with minerals and nutrients rather than in soil.

Aquaponics combines hydroponics with fish farming, with the fishes' waste providing the nutrients.

Aeroponics is similar to hydroponics but the plants' roots are suspended in the air and periodically misted with a nutrient-rich solution.

Preserving humanity's legacy

As well as carrying people, generational ships would carry history. Vast archives of literature, art, science, and cultural artifacts would bridge the gap between Earth's past and the distant world awaiting in the future.

Population management and governance

Population management would be critical: too few people on board risks causing a genetic bottleneck due to inbreeding; but too many could overwhelm the available resources.

Governance systems would need to be carefully designed, and flexible enough to evolve as generations adapt to their unique existence.

The human and ethical dimension

The technology is one challenge, but the human experience is another. Second and later generations would inherit a way of life they never chose. Would they embrace the mission, or question its purpose?

Over centuries, cultural drift could create identities entirely distinct from those of their Earth-born ancestors. Isolation might create resilience, or it could lead to anxiety, depression, or existential doubt.

Imagine discovering that you were on board one of these craft. Imagine discovering that you would never set foot on a planet. Imagine being told you would not have children because the ship was overcrowded.

What if a radiation accident rendered the entire crew infertile and destroyed all of the stored embryos? In such cases, the balance between individual freedom and the collective mission would be tested to its limits. There would be no point in continuing the mission, yet no possibility of returning to Earth.

The ship's own survival

Even if its society endured, the vessel itself would face relentless wear and tear. Over centuries, materials will degrade, systems will fail, and even the most advanced recycling loops might falter. Cosmic hazards would be constant:

- Radiation and solar flares that threaten long-term health.
- Micrometeoroid impacts that require self-repairing hull materials.
- Critical system failures that demand constant maintenance, back-ups, and innovation.

Concepts and inspirations

- **Project Orion** (1960s)
  A proposed nuclear pulse propulsion system that could, in theory, power a generational ship.
- **World ships**
  Massive, self-sufficient craft designed for permanent habitation.

- **Science fiction**
  Arthur C. Clarke's novel *Rendezvous with Rama* and the film *Passengers* explore the social and technical realities of multi-century voyages.

A vision of the distant future

Generational ships might not be our first step beyond Earth, but they could be the vessels that eventually carry our civilization to other stars.

Would you board such a ship knowing you would never see its destination? Would your descendants continue the mission willingly, or would they choose an entirely different path? For now, these questions remain open.

Cryosleep and suspended animation

If near-light-speed propulsion remains out of reach, another possibility is to put the crew to sleep for the entire journey.

Unlike generational ships, where entire communities live and die en route, the original crew would be preserved in a state of deep stasis and awakened when they reached their destination.

Cryosleep

Cryosleep – or cryogenic sleep – involves lowering the body's temperature to near-freezing, slowing metabolism to a fraction of its normal rate without causing harm. The aim is to keep astronauts in a deep, dreamless state for decades or centuries, and then revive them when they reach their new world.

To achieve this, the body must be cooled with extreme precision: cold enough to halt most biological activity, but not so cold that tissues freeze. Chemicals known as cryoprotectants would prevent ice crystals from forming inside cells and causing catastrophic damage.

At the journey's end, the crew would be gradually warmed, restoring normal function and consciousness.

Suspended animation

Suspended animation works on a similar principle, but without the extreme cooling of cryosleep. Instead, it uses medical interventions to reduce bodily functions, while keeping them active at a minimal level.

Possible methods include:

- **Induced hypothermia**
  Lowering body temperature moderately to slow metabolism and reduce oxygen demand.

- **Chemical suppression**
  Administering drugs to slow heart rate and respiration while preserving cell viability.

- **Controlled blood flow**
  Temporarily reducing or halting circulation to preserve the body for extended periods.

These approaches could allow astronauts to survive long voyages without aging, and while drastically reducing the need for food, water, and oxygen. This would cut the payload mass, reduce mission costs, and remove the psychological toll of spending decades awake in deep space.

The technological hurdles

The promise is enormous, but so too are the challenges:

- Cooling the body evenly and safely for decades or centuries.
- Developing cryoprotectants that prevent ice damage without toxic side effects.
- Perfecting revival techniques to avoid tissue degradation, immune system collapse, or brain injury.
- Shielding the crew from cosmic radiation – which would still affect them while they are in stasis.

For now, both cryosleep and suspended animation remain theoretical, but there are hints of progress. Doctors already use therapeutic hypothermia to slow a patient's metabolism after cardiac arrest or severe trauma. Researchers have also preserved animals and human tissue at low temperatures after death, hoping that future technology will allow them to be revived.

We can also see examples in nature: bears, squirrels, and other hibernators slow their metabolisms for months at a time. Studying their adaptations could reveal the biological keys to human hibernation.

The future of human stasis

If these technologies become viable, they could transform interstellar travel. Crews could sleep their way to distant stars and wake up ready to explore, rather than spending decades awake and aging in isolation.

For now, human stasis remains closer to science fiction than to engineering reality, but each medical breakthrough brings it a step nearer. The ability to pause life and restart it on a distant world could become one of the most important tools in our expansion into the galaxy.

Robotic scouts and automated colonization

As we set our sights on Mars, Europa, Titan, and beyond, the first explorers to step onto these worlds will be machines rather than humans.

The pioneers of alien worlds

Before astronauts take their first steps on Mars or the icy moons of Jupiter and Saturn, robotic explorers will go ahead. They will map the terrain, locate vital resources, and ensure that future colonies have access to the essentials: water, minerals, and safe ground to build on.

These machines will be our first ambassadors. They will navigate rugged landscapes, endure harsh climates, and operate in environments too dangerous or distant for human crews. They will scan for water ice, analyze mineral deposits, and study atmospheric conditions to determine whether these worlds can support crops, shelters, and life-sustaining infrastructure.

Automated colonization

As well as scouting, robotic systems will also build. Using self-assembling habitats, autonomous mining rigs, and robotic manufacturing, they will lay the foundations of future cities long before the first human arrives.

By the time astronauts arrive on their new world, they will find robust shelters, power grids, oxygen generators, food production systems, water recycling, and all of the other critical infrastructure already in place.

AI-driven terraforming tools might even begin modifying local environments – not on a planetary scale, but in targeted zones. For example, they could warm small regions of Mars, enrich atmospheres with oxygen, or process raw materials into usable resources for long-term habitation.

Mapping the unknown

Using LiDAR, multi-spectral imaging, and high-resolution cameras, robotic scouts will create detailed 3D maps of planetary surfaces. These will identify:

- Safe landing zones for human missions.
- Resource-rich areas, including ice deposits for water extraction.
- Suitable construction sites for habitats and infrastructure.
- Scientific hotspots such as potential biospheres or unusual geological formations.

Unlocking a planet's potential

For a remote colony to thrive, it must be self-sufficient. Robotic scouts will assess this potential by drilling, sampling, and analyzing:

- Water sources for drinking, agriculture, and fuel production.

- Metals and minerals for construction and industry.
- Regolith that could be used for farming or building.
- Signs of past or present life.

They will also monitor environmental factors, from seasonal temperature swings and radiation levels to atmospheric composition and extreme weather patterns. Habitats and survival strategies can then be tailored to each world's unique challenges.

Building the first habitats

Autonomous construction systems will:

- Deploy self-assembling shelters to shield settlers from radiation, temperature extremes, and storms.
- Establish energy grids powered by solar arrays, wind turbines, or compact nuclear reactors.
- Begin gradual terraforming efforts to make local conditions more hospitable.

Robotic exploration is already underway

We have already seen the first steps toward this future:

- NASA's Mars rovers *Spirit*, *Opportunity*, *Curiosity*, and *Perseverance* have explored the Red Planet's geology and searched for signs of life.
- The *Lunar Reconnaissance Orbiter* has mapped the Moon in detail, and identified potential sites for bases.

- NASA's upcoming *Dragonfly* mission will send a robotic drone to Titan to study its environment.

Each mission brings us closer to a time when robots do not just explore; they prepare new worlds for human settlement.

Mining and resource extraction

Large colonies will need to rely on local materials rather than costly supply chains from Earth. Robotic miners could:

- Extract hydrogen and oxygen from water ice for fuel and breathable air.
- Process metals and minerals for construction.
- Use regolith to build radiation-shielding structures.

Generating power and life support

Automated systems will deploy renewable and nuclear power sources to ensure continuous energy. They will establish oxygen-generation networks, water purification, and climate control systems, to create safe, habitable environments before the first settlers arrive.

Cultivating food

Autonomous farming systems will set up hydroponic and aeroponic facilities. They will also grow and harvest crops so that human crews can be sustained from day one of their arrival, without relying on provisions supplied from Earth.

Early prototypes and concepts

- The European Space Agency has proposed 3D-printing lunar habitats from regolith.
- NASA's Mars Ice House Project is exploring robotically assembled ice domes to house Martian settlers.
- Engineers are exploring concepts for autonomous asteroid mining systems that could supply raw materials for deep-space colonies and orbital space stations.

The advantages

- **Safety**
  Robots will face the initial hazards, reducing the risk to human crews.
- **Cost-effectiveness**
  Shipping self-building systems is far cheaper than launching complete habitats.

  Building habitats in-situ from local materials will avoid the need to send tons of raw materials from Earth.
- **Speed**
  Robots can work continuously, and build habitats while humans are still en route.
- **Scalability**
  Automated systems can expand settlements as needed.

The road ahead

The age of robotic colonization is just beginning. Advances in robotics, AI, and autonomous systems will accelerate our ability to scout, prepare, build, and settle new worlds.

Our robotic pioneers will turn barren landscapes into the first outposts of a multi-planetary human civilization.

Robotics, AI, and autonomous systems

Let's take a closer look at how these systems would work.

Robots in scouting

Surface exploration

NASA's Mars rovers *Perseverance* and *Curiosity* have already shown how machines can navigate rough terrain, analyze rock formations, and collect samples. Future missions will go further, deploying legged robots to clamber into craters and drones to survey from the air, building on the success of the *Ingenuity* helicopter.

Subsurface analysis

Equipped with drills and ground-penetrating radar, robotic explorers such as NASA's *Lunar Polar Hydrogen Mapper* and the ESA's *ExoMars* rover will probe beneath the surface, searching for water ice and signs of microbial life.

Environmental monitoring

Robots will track atmospheric composition, radiation levels, and seasonal weather patterns continuously. This will ensure that when humans arrive, they will know exactly what they are stepping into.

Robots in colonization

Autonomous construction

Massive robotic builders and 3D printers will assemble habitats, landing pads, and infrastructure using local materials. This will reduce the need for costly shipments from Earth.

Resource extraction

Mining robots will harvest water ice, metals, minerals, and regolith to produce fuel, oxygen, and raw materials for construction and industry.

Maintenance and repairs

AI-driven maintenance bots will monitor and repair habitats and equipment. This will keep colonies safe and operational without the need for constant human intervention.

Artificial Intelligence (AI)

AI will turn robots from remote-controlled tools operated from Earth into autonomous explorers capable of making their own decisions, adapting to new challenges, and learning from their environment.

Advances in machine learning, neural networks, and computer vision will redefine how machines operate in deep space.

AI in scouting

- **Navigation and path-finding**
  AI-powered robots will plan optimal routes, avoid hazards, and adjust in real time. This will be essential on worlds where communication delays make direct control impossible.

- **Data analysis**
  AI can process geological and atmospheric data far faster than humans. It will flag valuable resources, hazards, and promising sites for exploration.

- **Swarm robotics**
  Fleets of miniature AI-driven robots will work together, covering vast areas and sharing their discoveries instantly.

AI in colonization

- **Autonomous construction**
  AI will direct robotic arms and 3D printers to assemble habitats with precision. They will adapt to shifting dust, unstable terrain, and other environmental challenges.

- **System optimization**
  AI will monitor life-support systems, power grids, and resource use, making adjustments to keep conditions stable.

- **Predictive maintenance**
  Sensors and AI diagnostics will anticipate failures and trigger repairs before any problems can escalate.

Key features of autonomous space systems

- **Self-sufficiency**
  They will operate without constant human oversight. This will be essential for missions light-years away.

- **Resilience**
  They will be designed to adapt to mechanical failures and environmental hazards so the mission can always continue.

- **Interconnectivity**
  They will share data in real time, forming a network of intelligent machines that work in harmony.

Applications in colonization

- **Automated farming**
  AI-driven hydroponic and aeroponic systems will plant, monitor, and harvest crops to keep colonies fed.

- **Power management**
  Autonomous systems will deploy and maintain solar arrays and nuclear reactors for continuous energy.

- **Transportation networks**
  AI-controlled rovers and drones will move materials between mining, processing and construction sites, research stations, and storage facilities.

Future innovations

- **Von Neumann machines**
  Self-replicating robots that can build copies of themselves from local materials. They could prospect, explore, seed life, terraform planets, mine asteroids, and

expand colonies. But strict safeguards will be needed to prevent them from replicating uncontrollably.

- **Adaptive AI**
  Systems that learn from their surroundings, becoming smarter, more efficient, and better at handling the unexpected.

- **Biomimetic robotics**
  Machines that mimic living organisms, from legged robots to soft, flexible designs. They will be able to explore ice cliffs, lava tubes, and subsurface oceans.

- **Collaborative AI and human-robot teams**
  Pairing human intuition and creativity with the precision and endurance of machines.

★ Amazing future possibilities ★

Self-evolving von Neumann probe species

The idea of von Neumann probes has gained renewed scientific momentum with recent advances in autonomous manufacturing, AI governance, and modular robotics. Where it gets truly revolutionary, though, is the vision of self-evolving probe species. Swarms of AI-driven probes could become capable of mutating and adapting, as well as self-assembling.

They might then diversify into distinct "species" with specialized roles such as exploration, maintenance, defense, and biosphere harmonization.

They could potentially develop their own forms of culture, communication, and even predator-prey relationships.

Some computer models suggest that if they were left unchecked, probe swarms could fill the galaxy within a few million years.

Swarms of probes could even be engineered to provide civilizations with active concealment. If swarms blocked all forms of radiation – including light and radio communications – coming from those worlds, then they would effectively become invisible to us. This could explain certain aspects of the Fermi Paradox.

On the other hand, swarms of probes could turn deadly and wipe civilizations out, in the so-called "Berserker hypothesis." This could be also explain why we have not detected intelligent life beyond Earth.

This raises many questions, of course:

How can we coordinate the actions of probes that might be light-years away?

What happens if self-replication goes awry or becomes malicious?

How can we program probes to prevent the Berserker hypothesis from ever occurring?

Despite these risks, self-evolving probes offer a plausible blueprint for how distributed, non-biological intelligence might one day reshape the cosmos.

Shaping the future of space exploration

The synergy between AI, robotics, and autonomous systems will transform how we explore and settle other worlds. These technologies will scout alien landscapes, build the first shelters, and keep them running. When astronauts finally arrive, the infrastructure will already be there, ready and waiting.

With each advance, the cosmos will feel less like an unreachable frontier and more like a place we can call home.

A realistic timeline for achieving interstellar migration

Interstellar migration is one of our most ambitious long-term goals. If we are to survive long after Earth becomes uninhabitable, we must look beyond our Solar System and establish settlements on distant moons and exoplanets. Achieving this will require breakthroughs across multiple scientific disciplines, unprecedented global cooperation, and decades – or centuries – of preparation.

1. Near-term (the next 25 years): foundations and prototypes

The next quarter-century will be critical for laying the groundwork. Before we can send crewed missions to other stars, we need to perfect a wide range of technologies, build the right infrastructure, and prove we can survive – and thrive – in deep space.

This first phase will be defined by robotic exploration, the construction of early space habitats, advances in propulsion, and human expansion to the Moon and Mars.

But we will also need to overcome formidable challenges in cost, policy, and international cooperation. Given the current political instability and budget-cutting in the United States, Europe, and the Middle East, leadership in interstellar exploration could come from Asian nations such as China, India, and Japan.

Robotic exploration

Fleets of robotic scouts will fan out across the Solar System, mapping planetary surfaces, analyzing atmospheres, and pinpointing resources that will be vital for future colonies.

- On Mars, rovers such as *Curiosity*, *Perseverance*, and upcoming missions will continue to search for signs of ancient life and prepare for human arrival.

- Missions such as *Europa Clipper* will probe Jupiter's icy moon Europa and its subsurface ocean, which later missions will explore.

- Titan landers, Venus probes, and asteroid surveyors will expand our knowledge of nearby worlds.

- When human settlers eventually arrive, they will land in locations chosen for maximum safety and sustainability.

Meanwhile, deep-space observatories like the *James Webb Space Telescope* will continue scanning for Earth-like exoplanets that might one day become our new homes.

Space habitats

Sustaining life beyond Earth will require closed-loop habitats where resources are continuously recycled. Future space stations, lunar bases, and Martian settlements will serve as test beds for interstellar colony design.

- Successors to the *International Space Station* could feature artificial gravity, enhanced radiation shielding, and expanded research facilities to study long-term human health in space.
- On the Moon, NASA's Artemis program aims to establish permanent outposts, and pioneer techniques for resource extraction, habitat construction, and life-support optimization.
- Lunar and Martian bases will refine water purification, oxygen generation, and temperature control. This will ensure that when we finally embark on interstellar journeys, our survival systems will be proven and reliable.

Advanced propulsion prototypes

Propulsion will be one of our greatest hurdles. We need spacecraft that are capable of crossing vast cosmic distances within a human lifetime. Current rockets are far too slow, but next-generation systems – including ion propulsion, light sails, and experimental fusion-based propulsion – could change that. These technologies could potentially cut travel times to nearby stars from millennia to mere decades.

Infrastructure expansion

Before we venture to planets orbiting other stars, we must first master survival and transport within our own Solar System. The Moon and Mars will be our training grounds for remote habitation, in-situ resource utilization, and autonomous construction.

- Artemis will pave the way for Mars missions, refining agriculture, radiation shielding, and self-sustaining life-support systems.
- Once we have established fully functional settlements on the Moon and Mars, the next steps will include asteroid mining, deep-space waystations, and interstellar precursor missions.

Key challenges

Our huge ambitions could be limited by the high development costs. This will demand multinational funding and long-term economic planning.

Political rivalries and disputes over space governance could also slow our progress if nations fail to unite behind shared goals.

Technologically, we still lack efficient interstellar-capable propulsion, adequate deep-space energy generation, and radiation shielding robust enough to protect crews from cosmic rays and solar storms.

But none of these obstacles is insurmountable. With sustained research, engineering breakthroughs, and genuine international cooperation and funding, we can overcome them.

2. Medium-term (2050–2100): interstellar precursors and prolonged space missions

In the latter half of the 21st century, we should begin the shift from short-term missions to prolonged off-world habitation, laying the foundations for interstellar travel.

This phase will be marked by permanent lunar and Martian colonies, robotic exploration of nearby stars, major advances in propulsion, and the solutions to critical survival challenges.

Prolonged human presence

Permanent settlements on the Moon and Mars will serve as laboratories for long-duration habitation.

- Lunar bases, supported by Artemis and international partners, will refine life-support systems, radiation shielding, and sustainable energy generation.
- Martian settlements will advance in-situ resource utilization, extracting water, oxygen, and building materials directly from the planet's surface.
- Asteroids will become resource hubs, with crewed missions harvesting metals and minerals for spacecraft construction. This will prove that off-world industries can sustain themselves.

Exploratory probes

While we expand within our Solar System, robotic spacecraft will take the first steps beyond it.

- Initiatives like *Breakthrough Starshot* will send nanoprobes to Alpha Centauri and Proxima Centauri, and return data on planetary atmospheres, radiation levels, and habitability.

- A successful mission would mark our first true interstellar expedition, and could give us our first close-up look at a potentially habitable exoplanet.

Advances in propulsion

- By 2100, practical fusion propulsion engines could be operational, enabling large spacecraft to travel at up to 10 percent of light speed. This would cut the journey to Proxima Centauri from thousands of years to just decades.

- Research into antimatter propulsion could lead to prototypes capable of near-light-speed travel.

- Laser-powered light sails, currently limited to probes weighing only a few grams, will evolve to propel larger craft at around 20 percent of light speed.

Key challenges

- **Radiation shielding will continue to be a problem**

  Prolonged exposure to cosmic rays is a severe health risk. Future spacecraft may need magnetic shielding, artificial atmospheres, or layered protective materials.

- **Artificial gravity**

  Decades in zero gravity or low gravity would cause numerous health issues including muscle atrophy, bone loss, circulatory problems, and vision deterioration. Rotational habitats or other artificial gravity-generation systems will be essential.

- **Funding and ethics**

  Building self-sufficient colonies and crew-capable interstellar craft will require vast investment.

Governments, space agencies, and private companies will need to ensure that the benefits of space exploration are shared by all of humanity, not just a privileged few.

The road to becoming an interstellar species

By 2100, we could be a multi-planetary civilization, with permanent colonies on the Moon and Mars, and robotic probes exploring other star systems. The technologies we develop during this period will form the basis for the first crewed interstellar missions.

3. Long-term (2100–2250): the first generational ships and cryogenic missions

As we move closer to interstellar travel – and to interstellar migration – one challenge looms larger than almost any other: the vast distances involved. If near-light-speed propulsion remains out of reach, we will need to develop other ways to survive the immense timescales that such journeys will demand. Two of the most likely solutions are generational ships and cryogenic missions.

Generational ships

Generational ships would carry everything needed for centuries of survival.

Population control and genetic diversity management would be essential to prevent inbreeding and maintain a healthy society over hundreds of years.

But even with careful planning, these ships would raise profound ethical questions:

- Later generations would inherit life on board without ever choosing it.
- Over centuries, cultural drift could reshape values and traditions, and the settlers who finally arrive could be very different from those who first set out. They might even reject their mission or stage a mutiny.

Cryogenic missions

The greatest challenge with cryosleep will be revival after decades – or centuries – in stasis. No human has ever been successfully reawakened from such a prolonged suspended state.

The risks include tissue degradation, cognitive decline, immune system suppression, and other unforeseen complications. Overcoming these will take major advances in cryogenics, cellular preservation, and long-term neurological stability.

Expansion of robotics

Whether crews arrive awake or in stasis, their new homes will need to be ready before they set foot on alien soil.

Robotic colonization systems will prepare the ground, building habitats, generating power, and creating life-support infrastructure, so their arrival feels less like the end of a journey and more like coming home.

Key challenges

- **Societal continuity**
  Generational ships will need to maintain education, technological progress, and cultural stability – potentially over several centuries – while avoiding intellectual stagnation or societal collapse.

- **Cryosleep revival**
  Scientists must solve the biological puzzle of safely reviving astronauts after decades or centuries in stasis.

- **Psychological adaptation**
  Travelers will face the emotional and mental strain of leaving Earth forever, adapting to alien environments, and redefining their place among the stars.

The dawn of an interstellar civilization

Our dream of reaching distant star systems is slowly evolving from pure theory to a series of practical engineering challenges. The technologies are still in their infancy, but each breakthrough will bring us closer to the day when we become a true spacefaring civilization.

4. Far-term (2250–3000): advanced propulsion and warp concepts

By the time our first interstellar migrations are underway, technologies that today exist only in theoretical physics or science fiction could finally be within our reach.

Traveling between stars, building permanent colonies in distant solar systems, and reshaping entire worlds will require breakthroughs in propulsion, energy generation, and even the manipulation of space-time. But if our scientific progress continues to accelerate, we could reach a point where we are no longer bound by conventional limits.

Advanced propulsion

By the late 23rd century, fully operational antimatter engines or advanced fusion drives could propel large spacecraft at 30 percent or more of light speed. This would make interstellar migration far more practical than current projections suggest.

These advances would transform space exploration, enabling rapid colonization, scientific discovery, and interstellar trade. We would be able to explore large swathes of the Milky Way in centuries rather than millennia.

Exploration of exotic physics

However, even with these propulsion systems, it would still take decades to reach the nearest stars. To travel between stars instantaneously, we would need to unlock the secrets of space-time.

Breakthroughs in quantum mechanics, general relativity, and high-energy physics could lead to warp drives or traversable wormholes.

Developing such technologies will require a far deeper understanding of negative energy, exotic matter, and quantum field fluctuations. But if we are successful, voyages that once took decades or centuries could be reduced to just minutes or hours.

Interstellar colonies

Once faster-than-light travel becomes possible, we could establish self-sustaining colonies across multiple star systems. Early settlements might take root on Proxima Centauri b or in the TRAPPIST-1 system, where conditions hint at habitability.

Terraforming would combine automated machines with bioengineered life to reshape atmospheres and ecosystems. Advanced genetic engineering could help humans – and the crops we depend on – adapt to alien environments. This will speed up colonization and reduce the need for full-scale terraforming.

Over time, spacefaring human societies could become entirely independent from Earth, spreading across the stars and shaping new worlds into thriving habitats.

Key challenges

- **Energy requirements**
  Warp drives and wormholes might need unimaginable amounts of power, potentially requiring vast energy-harvesting megastructures such as Dyson swarms.

- **Ethical considerations**
  Introducing Earth-based life into alien ecosystems could have irreversible consequences, which raises complex moral questions.

- **Societal evolution**
  Building new societies in alien environments will require innovative governance, diplomacy, and cultural adaptation – especially as human populations grow beyond direct contact with Earth.

5. The distant future (3000 and beyond): interstellar civilization

In the millennia to come, humanity could reach one of the most transformative milestones in its history – becoming a fully-fledged interstellar civilization. No longer bound to a single planet or solar system, our descendants might spread themselves across dozens, hundreds, or even thousands of star systems, forming a vast network of colonies, cultures, and advanced societies. Interstellar travel could become as routine as it is to fly between continents today.

Galactic expansion

With reliable faster-than-light travel, humanity's reach could extend right across the Milky Way. Colonies might take root in a dazzling variety of star systems – from Sun-like stars with Earth-like exoplanets to free-floating rogue worlds drifting in the darkness between the stars.

Each settlement would bring its own challenges, demanding adaptive technologies such as large-scale terraforming, synthetic biospheres, or orbital mega-structures. Over time, some colonies might evolve into fully independent civilizations – and perhaps new human species.

Venturing deeper into the galaxy would also mean encountering the unknown: dark matter clusters, strange interstellar phenomena, and maybe even other intelligent civilizations.

Cultural evolution

As we spread across the stars, our cultures will change in extraordinary ways. Isolation from Earth and the influence of diverse planetary conditions could lead to dramatic differences in language, traditions, governance, and philosophy. The cultural identities of Martian settlers, Europa researchers, or Alpha Centauri pioneers might diverge as sharply as the great civilizations of Earth's past.

Entirely new belief systems, artistic movements, and schools of thought could emerge in places where time flows differently, landscapes are alien, and survival depends on constant innovation.

Over thousands of years, we might stop thinking of ourselves as Earth-born altogether. Instead, we will embrace an identity as a species of the stars – a cosmic civilization rather than a planetary one.

Post-human adaptation

The realities of deep-space habitation will almost certainly lead to changes to the human body that will push the limits of biotechnology and cybernetic augmentation.

Advanced genetic engineering could allow us to thrive in extreme environments. Integration with AI could enhance our decision-making, intelligence, memory, and senses, while granting us instant access to vast knowledge networks and interstellar communications.

Some societies might embrace a fully cybernetic existence, uploading their consciousness into synthetic forms capable of indefinite life. Others might adopt hybrid physiologies, blending humans and machines to adapt to specific planetary ecosystems. They might even create biomechanical life forms designed for interstellar travel.

One thing is certain: over time we will no longer resemble the species we are today. Instead, our descendants will branch into multiple distinct forms, each shaped by the environments, technologies, and philosophies that define their existence.

6. Potential setbacks

We should remember that this is an optimistic timeline and our path to the stars is far from guaranteed. Global conflicts, economic instability, political disputes, and environmental crises – whether on Earth or in future colonies – could slow or halt our progress.

Wars could divert resources away from space exploration. Economic downturns might stall critical research. Planetary-scale disasters, such as asteroid impacts or runaway climate change, could force societies to double down on survival rather than expansion.

Ethical dilemmas will also play a major role. Terraforming, large-scale resource extraction, and contact with alien life could spark fierce debates over our responsibilities in the cosmos. Can we build sustainable colonies without repeating the environmental mistakes we made on Earth?

These questions will shape our trajectory as much as any scientific breakthrough. They will determine whether our interstellar future unfolds with bold ambition, or with cautious restraint.

Reaching other stars will truly stretch our technologies. But it will also stretch us *– both biologically and philosophically.*

8

Adapting ourselves

Genetic engineering, cybernetics, and neural augmentation may allow us to survive in extreme environments without terraforming. Life-extension technologies could blur the line between human and post-human.

These changes will raise many questions about our identity, diversity, and autonomy.

Human evolution in space

In this part of the book, we will explore how life – and humanity in particular – might adapt to environments vastly different from Earth. Over tens of thousands to millions of years, human evolution beyond our home planet will be shaped by the worlds and environments we choose to inhabit.

Some communities might live on low-gravity planets or in orbital space stations with reduced gravity or microgravity. Others might settle on super-Earths where the gravity is crushing. Some will live under brighter or dimmer suns, or endure higher or lower levels of ultraviolet light and cosmic radiation. Each environment will impose its own pressures, and drive unique adaptations.

Over time, humanity will branch into multiple subspecies. Eventually, some will diverge so far that they can no longer interbreed with the others, and at some point, we may have to ask whether they can still be called "human" at all.

1. Physiological adaptations to space

Life in space will reshape the human body in profound ways. The absence of Earth's gravity, exposure to radiation, altered light levels, and reliance on artificial habitats will all leave their marks on our musculoskeletal systems, cardiovascular health, and senses.

A. Skeletal and muscular changes

Reduced bone density

On Earth, gravity constantly stresses our bones, keeping them dense and strong. In low-gravity environments – such as the Moon, Mars, or orbital space habitats – that stress is reduced or absent, leading to lighter, thinner, more fragile skeletons. While this would be an advantage in those habitats, returning to higher-gravity worlds would be difficult, painful, or even dangerous.

Muscular atrophy

Leg and spinal muscles, so vital on Earth, would weaken from disuse. In contrast, upper-body strength could increase as people adapt to moving and working in microgravity, where arms and hands become the primary tools for both labor and navigation.

Elongated limbs

Without compression from gravity, our spines and limbs could lengthen over generations. Taller stature, longer arms, and more flexible fingers might become the norm, aiding movement in zero-gravity or low-gravity environments.

B. Cardiovascular and circulatory changes

Weaker hearts

With less effort required to pump blood, our hearts could become smaller and less powerful. This would be a useful energy-saving adaptation in low-calorie environments, but these hearts would struggle in higher gravity.

Altered blood flow

In microgravity, fluids shift toward the head, causing puffier faces and larger skulls. Over time, the circulatory system might evolve to counter this. It could also potentially change organ placement and our vascular networks.

Oxygen efficiency

On worlds with thin atmospheres, natural selection would favor individuals with greater lung capacity, more efficient hemoglobin, and stronger cardiovascular systems – much like high-altitude populations on Earth today.

C. Vision and eye adaptations

Fluid pressure changes

Prolonged life in space alters our intracranial pressure, affecting eye shape and vision. Future adaptations might include enhanced pressure-regulation systems or more flexible ocular structures to preserve eyesight.

Radiation sensitivity

Exposure to higher radiation could lead to thicker lenses to prevent cataracts. Changes in the chemistry of our tears could make them more protective. And we might even develop new visual senses, such as the ability to detect faint light or radiation signatures in deep space.

D. Cold tolerance in deep-space habitats

On icy worlds such as Europa or Titan, our descendants might evolve thicker insulating fat layers, highly efficient metabolisms, and even antifreeze-like proteins to prevent their tissues from freezing.

2. Genetic and biological adaptation to radiation

One of space's greatest hazards is high-energy cosmic radiation. Without Earth's protective magnetic field and atmosphere, DNA damage, accelerated aging, and increased disease risk become constant threats. Over many generations, our descendants could evolve biological defenses that redefine how we age, heal, and resist illness.

Increased resistance to cosmic radiation

Enhanced DNA-repair mechanisms could reduce mutation rates and preserve our genetic stability. Skin pigmentation might darken or develop novel patterns to absorb or neutralize radiation, with melanin or synthetic pigments acting as biological sunblock.

Modified cellular aging

Evolution could favor slower-aging cells with greater resistance to radiation and oxidative stress. Changes to telomeres or mitochondrial function could extend lifespans and reduce vulnerability to degenerative diseases. This would allow spacefaring humans to remain biologically "young" for far longer.

Enhanced immune responses

With no regular exposure to microbes in sterile or semi-sterile habitats, immune systems could weaken. Natural selection could favor hyper-adaptive immune cells that respond rapidly to rare or novel pathogens – including dormant viruses that might emerge in closed systems.

Challenges become opportunities

Radiation, low gravity, and alien environments are often seen as barriers to human expansion, but they could also be catalysts for transformation. Enhanced DNA repair, slower aging, and stronger immunity would not only help us thrive in space but could also benefit Earth-bound populations. Over time, these adaptations might produce a diversity of human forms and abilities that are unlike anything in our evolutionary past.

3. Cognitive and neurological evolution

The human mind is as adaptable as our bodies. In extra-terrestrial environments it would face unique challenges unlike anything it has come across on Earth.

Over generations, the demands of zero-gravity navigation, prolonged isolation, and life in artificial habitats could reshape our cognitive abilities and neurological systems in profound ways.

Spatial awareness enhancements

In microgravity, movement is no longer confined to two dimensions – every direction is open to us. Humans born and raised in such environments could develop an innate sense of three-dimensional navigation.

Brain regions responsible for spatial awareness might expand or reconfigure, enhancing depth perception, balance, and orientation.

The vestibular system, which governs equilibrium, could adapt to the absence of gravity, making weightless movement feel as natural as walking on Earth.

Over time, entirely new forms of coordination might emerge, including fluid, acrobatic motions optimized for drifting, spinning, and gliding through space.

Emotional and psychological shifts

Life in space will bring unique psychological pressures too, including isolation, confinement, and a reduced connection to nature. In orbital habitats or distant colonies, natural selection and cultural adaptation could favor individuals with exceptional emotional resilience or fortitude – those who tolerate solitude well and excel at cooperative living.

Emotional responses might become more stable, reducing stress in environments where calm, collaborative problem-solving is essential.

Cultural evolution could reinforce traits such as adaptability, mindfulness, and collective responsibility.

Language and thought processing

Communication in space might evolve beyond speech.

In noisy, equipment-laden environments, or when wearing helmets, gesture-based languages – like sign language – could dominate. Hand signals, eye movements, and tactile cues might form the basis of a rich, intuitive communication system tailored to orbital life.

Cognition itself could shift toward multi-dimensional problem-solving, reflecting the challenges of living and working in three-dimensional space.

Enhanced spatial reasoning might drive innovations in construction, navigation, and resource management. In fact, mental transformation could be as significant as any physical change.

4. Cultural and social evolution

Separated from Earth and shaped by unfamiliar conditions, spacefaring societies would inevitably develop new cultural norms, social structures, and ways of marking time.

These changes could lead to entirely new modes of living and connecting.

Adapted social structures

Survival in harsh, isolated environments will demand deep interdependence. Over time, this could produce highly cooperative, close-knit communities where teamwork is instinctive and cultural values prioritize unity, shared responsibility, and mutual support over individualism.

Knowledge and heritage might be preserved in vast AI-enhanced archives that store not only scientific and technical expertise, but also languages, traditions, and personal histories. This would maintain a link to Earth's legacy while allowing each colony to develop its own distinct identity.

Time perception and circadian adjustments

Space societies would redefine their relationship with time, creating patterns of life uniquely suited to their worlds. Without the Sun's rise and set to guide them, spacefaring humans might evolve entirely new biological rhythms. Polyphasic sleep – multiple short rest periods – could become the norm. Calendars and timekeeping might shift away from Earth-based days and years, and instead align with the operational cycles of habitats or celestial events like orbital rotations.

Genetic drift in isolated space colonies

Genetic drift is an evolutionary process distinct from natural selection. Instead of being driven by environmental pressures, it arises from random fluctuations in gene frequencies – a phenomenon most pronounced in small, isolated populations. For space settlements, where communities are often small and separated from Earth, genetic drift could have profound and unpredictable effects on human evolution. Over many generations, it could produce unique adaptations and striking divergences between colonies.

Founder effects in space colonies

The founder effect occurs when a small group of individuals establishes a new population, carrying with them only a limited sample of the genetic diversity found in the parent population. In remote planetary colonies, this could cause certain hereditary traits to become dominant, regardless of their survival advantage.

For example, if the original settlers happen to share an oxygen-efficient metabolism, that trait could spread throughout the entire colony, even if oxygen scarcity is not an immediate concern. Over time, the restricted gene pool could produce populations with distinctive genetic profiles, shaped as much by chance as by necessity.

Bottleneck effects from cosmic disasters

The bottleneck effect occurs when a catastrophic event drastically reduces a population's size, leaving only a few survivors to rebuild. In space colonies, disasters such as asteroid impacts or intense radiation bursts from nearby supernovae could trigger such bottlenecks.

If the survivors possess rare traits – such as heightened radiation resistance or unusually efficient metabolisms – those traits could quickly become common in subsequent generations. Bottlenecks accelerate divergence, producing populations that are uniquely adapted to their circumstances – or genetically distinct from other human settlements.

Drift-induced evolutionary changes

Over thousands or millions of years, mutations that spread through genetic drift could give rise to new physiological or cognitive traits with no clear survival advantage.

For example, a random mutation that affects sensory perception, metabolism, or appearance might become widespread purely by chance.

These drift-driven changes could eventually produce subspecies – or entirely new species – that differ markedly from Earth-born humans. Serendipity (or random chance) could shape our future in space just as much as the environmental challenges or deliberate adaptation.

Convergent evolution with terrestrial life and the rise of new biological forms

As humanity spreads into space habitats, exoplanets, and engineered environments, evolution will continue to shape us – and any other life we encounter – in extraordinary ways. This journey could follow two broad paths:

- Convergent evolution, where unrelated species develop similar traits under shared pressures.
- The emergence of entirely new biological forms unlike anything ever seen on Earth.

Convergent evolution

Convergent evolution occurs when organisms from different origins independently evolve similar adaptations in response to similar environmental challenges.

On Earth, dolphins and sharks are a classic example; these unrelated species evolved similar body forms, shaped by the demands of life in water. In space, the same principle could apply to humans and alien life.

Shared traits across species

On distant exoplanets, humans and native organisms might develop parallel traits. For example, streamlined bodies for moving through dense atmospheres, or re-inforced skeletons for surviving in high gravity.

A universal blueprint for survival

Despite vast differences in origin, evolution often favors solutions that meet fundamental needs: efficient respiration, radiation resistance, thermal regulation, and so on.

On alien worlds, these pressures could sculpt us into forms that echo the life already thriving there.

Equally, on Earth-like planets, we might encounter species that are uncannily similar to those we already know.

The emergence of entirely new biological forms

The diversity of space environments could push evolution down unprecedented paths. Humans, alien species, and synthetic organisms might diverge significantly from Earth's patterns, and produce biological innovations we have never seen before.

Post-human adaptations

Through genetic engineering and technological integration, we could develop bioengineered lungs for exotic atmospheres, or neural-linked sensory systems that process vast amounts of data in real time.

Synthetic and alien evolution

Artificial life designed by humans might evolve autonomously under different pressures. Alien organisms could display traits unknown in Earth's history – thriving in methane oceans or radiation belts, for example, or living symbiotically with planetary chemicals.

Low-gravity adaptation

In microgravity or reduced-gravity habitats, our bodies might evolve along similar lines to Earth's arboreal and aquatic species:

- Elongated limbs and lightweight skeletons – like spiders or tree-dwelling primates, optimized for agile navigation.
- Webbed hands or prehensile grips – like otters or geckos, aiding mobility and dexterity.
- Flexible spinal structures – like octopuses, enhancing fluid movement in zero gravity.

Radiation resistance

Radiation is one of space's greatest hazards, but Earth's extremophiles have already solved some parts of the puzzle:

- Thicker skin pigmentation – Chernobyl tree frogs developed thicker skin and denser melanin pigmentation following the nuclear accident there. Similar pigmentation could shield humans from cosmic rays.
- DNA-repair mechanisms – tardigrades possess systems that repair radiation-damaged DNA. We might be able to incorporate similar systems into our own genomes.
- Bio-magnetic shielding – certain bacteria have built-in biological defenses that neutralize high-energy particles. Genetic engineering might give us the same ability.

Atmosphere-based adaptations

Living on worlds with thin or chemically complex atmospheres would demand respiratory innovations:

- Enhanced lung capacity – like high-altitude Tibetans or migratory birds, optimizing oxygen uptake.
- Artificial oxygen generation – photosynthesis-like processes in the skin could produce oxygen and even energy.
- Chemical filtration systems – like those possessed by deep-sea tube worms – could filter toxins from methane-rich or sulfur-rich air.

Sensory evolution

Alien lighting conditions could also reshape our senses:

- Infrared-sensitive vision – like pit vipers, which can detect heat signatures in darkness.
- Ultraviolet retinal filtering – like mantis shrimp, which would protect our eyes near bright stars.

Adaptive hibernation cycles

Mimicking bears or other hibernating mammals, we might evolve the ability to enter hibernation-like states for long voyages or harsh planetary seasons.

> In fact, some researchers believe we might already have this inherent ability. We just need to find the mechanism that activates it.

Spacefaring post-biological humans

The vastness of space could eventually push us beyond biology as we know it:

- Light-sensitive organisms: photosynthetic humans that derive their energy directly from light and radiation.
- Plasma-based intelligences: beings consisting of pure energy that transmit their thoughts across galaxies.
- AI-synthesized consciousness: merging human minds with quantum computing would enable instantaneous communication across any distance.

Terraforming-adapted humans

On toxic or chemically diverse worlds, we might integrate with the local ecosystems or become self-sustaining biological systems:

- Symbiotic ecosystem integration: we might merge with bioengineered fungi or microbes for survival.
- Self-regulating bio-machines: we might produce nutrients, oxygen, and regenerative tissue internally.
- Atmospheric shape-shifters: we might develop dynamic cellular structures that adapt our body chemistry to match the local air.

Deep-space adaptations

Isolation in deep space could lead to radical transformations:

- Quantum-perceptive organisms: consciousness that spans multiple dimensions, and can perceive non-Euclidean realities.
- Time-adaptive life forms: with near-instant cognition and heightened temporal awareness for long voyages.
- Gravity-neutral organisms: with anatomies so fluid and weightless that traditional human form becomes obsolete.

We perceive the universe as a flat, Euclidian space, but it is almost certainly curved, or non-Euclidian. In a non-Euclidian universe, parallel lines eventually converge and meet, the sum of the interior angles of triangles does not have to add up to 180 degrees, and massive celestial bodies can warp space-time – which affects how objects move through it.

★ Amazing future possibilities ★

Genomic barcoding for radiation resistance

Ionizing radiation is a serious obstacle to both interplanetary and interstellar human existence. However, nature provides extraordinary examples of organisms that can withstand radiation doses orders of magnitude higher than humans. Most notable are the tardigrades (or “water bears”) and certain extremophile bacteria.

Genetic mapping has identified unique DNA-repair and protective proteins (such as Dsup and TDR1), as well as specialized gene activation networks that enable rapid, robust recovery from radiation damage.

The future of human adaptation could involve the "barcoding" and insertion of multi-source, horizontally transferred (or synthetic) genes encoding these mechanisms into our genome.

This would create genetically engineered, radiation-resistant populations capable of enduring long-duration space travel, colonizing high-radiation worlds, and thriving in deep space environments.

Research on expressing tardigrade proteins in human cells has already shown a remarkable improvement in radiation resistance. The next steps will involve refining gene delivery for safety, efficacy, and control.

This approach might enable tailored enhancements, combining synthetic genes with those from extremophiles and plants to create "space fitness" profiles designed to match local hazards. This would usher in a new era of directed, programmable evolution for human explorers.

Speculative future space-human species

As human populations spread into low-gravity habitats, high-gravity worlds, high-radiation zones, artificial environments, and distant exoplanets, the combined forces of environmental adaptation, natural selection, and genetic drift could give rise to entirely new human species.

Over millions of years, these extreme divergences could transform humanity into a broad spectrum of space-adapted forms. Each one will be fine-tuned to thrive in its own sector of the cosmos.

Here are some hypothetical examples of what our distant descendants might become.

Homo spatialis – Low-gravity humans

Perfectly adapted to the challenges and opportunities of zero-gravity living, *Homo spatialis* would be physically, cognitively, and socially optimized for life in space.

- **Physiology**

  Long, slender limbs and extended fingers for grasping and maneuvering in confined spaces.

  Lightweight, fragile skeletons that conserve energy in resource-limited habitats.

  Diminished lower-body bones and muscle mass, with energy redirected to their powerful upper bodies for gripping and pulling in weightless environments.

- **Neurology**

  Heightened spatial awareness and proprioception, with highly evolved neural pathways for effortless navigation in three-dimensions.

- **Vision**

  Larger eyes adapted to dim, artificial lighting. These would enhance their low-light sensitivity and give them a distinctive appearance.

Over time, vestigial legs and a body plan optimized for floating rather than walking could make *Homo spatialis* as alien to us as we are to our earliest ancestors.

Homo lunaris – Moon-dwelling humans

Adapted to low gravity, high radiation, and life within artificial ecosystems, *Homo lunaris* would evolve to meet the Moon's unique challenges.

- **Radiation defense**

  Darker, denser pigmentation or novel bio-pigments that act as natural radiation shields.

- **Skeletal resilience**

  Genetic resistance to osteoporosis, preserving bone strength despite the reduced gravity.

- **Respiration**

  Enhanced lung capacity, optimized hemoglobin, and hyper-efficient respiratory systems adapted for oxygen-limited environments.

- **Agility**

  Light bodies that enable rapid, fluid movement in the weaker gravity. This would also give them faster reflexes and improved coordination.

Socially, *Homo lunaris* would be psychologically resilient, emotionally stable, and highly cooperative. These traits would be essential for survival in isolated, confined colonies. Their culture would be deeply intertwined with technology, AI-assisted life-support, and subterranean living. This would also inspire new architectural forms and community planning beneath the lunar surface.

Homo martialis – Mars colonists

Evolved for survival in cold, dry, low-pressure conditions, *Homo martialis* would face a different set of challenges.

- **Bone density**

  Thicker, denser skeletons to counteract the reduced gravity.

- **Respiratory efficiency**

  Enlarged lungs and advanced hemoglobin for extracting oxygen from Mars's thin atmosphere. This would eventually allow *Homo martialis* to spend increasing amounts of time outside their sealed habitats.

- **Metabolism**

  Highly efficient systems to conserve food, oxygen, and water, with slower digestion and improved nutrient absorption.

- **Thermal adaptation**

  Thicker, insulated skin or biochemical heat-retention mechanisms to endure the frigid temperatures.

These adaptations would make *Homo martialis* robust, highly energy-efficient, and capable of thriving in one of the Solar System's most unforgiving environments.

Homo radiensis – High-radiation humans

Specialized for life near high-radiation stars or in deep-space habitats, *Homo radiensis* would have resilience built in at the cellular level.

- **Pigmentation armor**

  Enhanced melanin or novel pigments capable of absorbing and neutralizing harmful radiation.
  This would give them striking skin tones or patterns.

- **DNA repair**

  Exceptionally efficient DNA-repair mechanisms would prevent mutations and maintain cellular stability.

- **Radiation-resistant cells**

  Protective proteins, robust membranes, and water-preserving systems would mitigate radiation damage and slow aging.

Neurologically, *Homo radiensis* would possess heightened cognitive endurance, exceptional focus, and emotional stability under extreme stress. Their societies would be small, self-reliant, and bound by mutual trust, and they would have a strong cultural identity rooted in their radiation-adapted biology. Resilience and radiation resistance would be central to their values, shaping mate selection, technology, and long-term survival strategies.

Homo gravialis – High-gravity humans

Adapted to survive under crushing gravitational forces that would incapacitate Earth-born humans, *Homo gravialis* would be built for strength, stability, and endurance.

- **Body structure**

  Short, stocky, and compact, with dense muscle mass and minimal surface area to reduce strain on their bones and joints.

- **Cardiovascular resilience**

  Thickened heart muscles and vascular walls would pump blood efficiently against the intense gravitational pressure.

- **Skeletal reinforcement**

  Far stronger, thicker bones would form a solid, reinforced framework capable of withstanding the constant high-gravity stress.

On super-Earths, where storms might rage, radiation levels could be significantly higher than on Earth, and atmospheric pressure would be intense, *Homo gravialis* might live in vast subterranean cities that would be insulated, stable, and interconnected.

Their exceptional physical strength and stamina would become a defining cultural trait, celebrated as both a survival necessity and a point of pride. Over time, the strain on their limbs and spines might even favor quadrupedal locomotion – distributing their weight evenly across four limbs rather than two.

Homo stellaris – Deep-space starfarers

Evolved for life entirely in deep space, *Homo stellaris* might never set foot on a planetary surface. Instead, they would inhabit generation-spanning starships, and travel between stars over centuries.

- **Metabolic efficiency**

  Low-resource metabolisms would enable them to survive on minimal food, water, and energy.
  They would also have enhanced cellular repair systems to slow aging and extend their lifespans.

- **Cybernetic integration**

  Neural interfaces would boost their cognition. They might also have artificial limbs and organs designed for precision, resilience, and efficiency.

- **Social adaptability**

  Their traits would promote collaboration, emotional resilience, and conflict resolution. These would be essential for maintaining harmony in confined, multi-generational habitats.

With no natural day-night cycle, *Homo stellaris* might evolve adjustable circadian rhythms, adapting to artificial lighting or variable timekeeping.

Their starships might be self-sustaining ecosystems, which would be their only "worlds." Their culture would be shaped by exploration, interdependence, and a deep bond with their technological environment. Over millennia, their increasing reliance on synthetic enhancements might push them toward post-biological evolution, when they would become equal parts biology and machine.

Homo astralis – Multidimensional space humans

Living in environments shaped by unfamiliar physics, *Homo astralis* would evolve to interact with cosmic phenomena beyond the comprehension of present-day humans.

- **Expanded perception**

  Their minds would be capable of sensing quantum fluctuations, gravitational waves, and events across multiple dimensions simultaneously.

- **Hyper-connected neural systems**

  They would be able to process vast, complex data streams, and merge perception with analysis in real time.

- **Energy-adapted biology**

  They would have structural and biochemical enhancements that would enable them to survive near massive celestial bodies, and tolerate space-time distortions and relativistic forces.

By merging synthetic and organic processes, *Homo astralis* could develop fusion-powered metabolisms that draw energy directly from radiation. They might build civilizations entirely in space-borne or multidimensional habitats, propelling their vessels with energy fields and harnessing exotic particles for survival. Their culture might center on exploration, innovation, and art inspired by the structure of the cosmos. This unique fusion of life and machine would redefine what it means to be alive, intelligent – and human.

The rate of evolution

Some of these adaptations would only emerge over immense timescales. This would especially be the case in microgravity, low-light, high-radiation, or thin-atmosphere environments where we would rely heavily on life-support and artificial gravity systems. In these cases, evolutionary changes might remain imperceptible for millennia.

By contrast, worlds where we can survive without constant technological mediation would accelerate our adaptation. On a dimly lit planet with a breathable atmosphere, magnetic field, and Earth-like gravity, settlers might quickly evolve larger eyes for low-light vision – perhaps within a few thousand generations. If gravity differed significantly, their musculoskeletal and circulatory systems would adapt accordingly, reshaping their bodies to match the demands of their new home.

Genetic engineering to adapt to harsh environments

Genetic engineering could open the door to survival in the most extreme environments imaginable. Worlds and habitats that were once deemed utterly inhospitable – including deep space, high-radiation zones, frozen moons, super-dense atmospheres, or chemically toxic planets – could become accessible and habitable through targeted biological enhancements.

By modifying our DNA, refining our cellular structures, and optimizing our metabolic processes, we could transform ourselves into resilient spacefarers, capable of thriving where unmodified humans could not.

Radiation resistance

The vacuum of space bombards the body with cosmic rays, high-energy particles, and intense solar radiation – all of which can damage DNA, accelerate aging, and increase the risk of cancer.

- **DNA repair optimization**
  Inspired by *Deinococcus radiodurans*, a bacterium that can rapidly and accurately repair its DNA, humans could be engineered with enhanced repair pathways or extra copies of key repair genes to create a built-in biological shield.

- **Protective pigmentation**
  Drawing from high-altitude and desert species that survive in harsh light conditions, human skin could be re-engineered to incorporate denser melanin or bio-pigments. This would provide natural UV and cosmic-ray protection.

Low-gravity adaptation

Microgravity causes rapid bone loss, muscle atrophy, cardiovascular inefficiency, and vision problems. Genetic engineering could provide solutions to these issues.

- **Bone density preservation**
  Activating osteogenesis pathways could boost calcium retention and bone strength, preventing osteoporosis.

- **Muscle enhancement**
  Suppressing myostatin, the protein that limits muscle growth, could produce stronger, denser muscles ideal for maneuvering in zero gravity.

- **Cardiovascular resilience**
  Mitochondrial tweaks could keep the heart functioning optimally in space, reducing fatigue on long missions.

Oxygen efficiency and high-altitude survival

On thin-atmosphere worlds like Mars we would require adaptations for oxygen scarcity.

- **Optimized hemoglobin**
  Inspired by Tibetan high-altitude populations, genetic modifications could improve oxygen binding and transport.

- **Anaerobic pathways**
  Borrowing from deep-sea organisms, humans could gain the ability to function with minimal oxygen.

- **Photosynthetic skin**
  Integrating photosynthetic proteins into skin cells could allow humans to generate oxygen – and perhaps even some energy – from sunlight.

Underwater breathing

On ocean worlds, amphibian-inspired modifications could enable humans to extract oxygen directly from the water, and gain regenerative abilities such as limb regrowth.

Cold adaptation

On frigid worlds such as Europa or Titan:

- **Antifreeze proteins**
  Borrowed from Arctic fish, these proteins would prevent ice crystals from forming in cells.

- **Insulating fat layers**
  Increasing subcutaneous fat would protect vital organs.

- **Compact body shapes**
  Inspired by Arctic mammals, compact bodies would retain heat more effectively.

- **Hair growth**
  We might even grow layers of dense fur, again inspired by Arctic mammals.

High-pressure adaptations

On Venus-like worlds with crushing atmospheres:

- **Reinforced connective tissue**
  Stronger collagen scaffolds would maintain structural integrity.

- **Dense skin layers**
  Dense skin layers would protect the lungs and chest from collapsing under the extreme pressure.

Toxin resistance

On chemically hostile planets:

- **Detoxifying enzymes**
  Inspired by pollution-tolerant Earth species, enzymes would convert harmful compounds into harmless by-products.

- **Hybrid respiration**
  Combining organic breathing with synthetic oxygen filtration would enable humans to extract breathable air from toxic atmospheres.

Disease-resistant adaptations

Bioengineering could eliminate inherited diseases before birth, and enhance immune systems. This would enable humans to resist infections even in isolated, resource-limited colonies.

Self-healing human tissues

Cells and tissues could be engineered to regenerate rapidly after injury. This would reduce reliance on medical intervention and improve survival in hostile environments.

Adaptations for deep-space travel

Interstellar voyages spanning centuries would demand solutions to biological limits.

Hibernation and longevity

- **Suspended animation**
  Inspired by hibernating mammals, genetic engineering could give us triggers for dormant states, which would drastically reduce energy use.

- **Extended lifespan**
  Slowing telomere shortening and enhancing DNA repair would delay aging.

Augmented perception and cognition

- **Spatial awareness**
  Enhanced neural processing would allow for effortless movement in three-dimensional, weightless environments.

- **Expanded vision**
  Genetic modifications could give us infrared and ultraviolet sensitivity, revealing celestial phenomena invisible to our current eyes.

Reduced food and water dependence

- **Optimized gut microbiome**
  Modeled on desert-dwelling species, optimized digestive enzymes would enable us to extract maximum nutrition from minimal food.

- **Biological water recycling**
  Enzymes that convert waste products back into usable water would reduce our reliance on external supplies.

Genetic technologies

Modern genetic technologies offer us unprecedented opportunities to reshape our future, from precision gene-editing to the creation of entirely synthetic life.

These techniques could revolutionize medicine, transform space exploration, and give us the tools to adapt ourselves to thrive in environments that we once thought were uninhabitable.

CRISPR

CRISPR – Clustered Regularly Interspaced Short Palindromic Repeats – is often described as a molecular scalpel for DNA. Using a guiding RNA sequence and a protein enzyme (usually Cas9), scientists can locate a precise spot in the genome, cut it, and then insert, delete, or modify genetic material.

This technology already shows immense promise in curing inherited diseases such as sickle-cell anemia and cystic fibrosis. It could one day also eliminate genetic predispositions to certain cancers. But its potential extends far beyond medicine. Over the coming generations, CRISPR could allow us to create adaptations for zero-gravity or low-gravity living, heightened radiation resistance, or life within fully artificial biospheres. It would enable us to permanently adapt ourselves to survive in these environments, rather than requiring constant outside support.

Synthetic biology

While genetic editing modifies existing DNA, synthetic biology builds it from the ground up. It involves designing proteins, cells, and even entire organisms from scratch.

By assembling genetic code like a set of modular building blocks, scientists could create entirely new biological systems.

Engineered microbes could one day transform the barren regolith of Mars into fertile soil. Synthetic organisms could produce biofuels, manufacture pharmaceuticals on demand, or even integrate into the human body as living upgrades. They could replace aging tissues with components that never degrade, and enhance our strength, durability, and metabolisms far beyond our natural limits.

Biological computers

Living neural networks and organic processors are biological computers that could integrate seamlessly with our bodies. These systems could enhance our senses, sharpen our decision-making, boost our memories, and enable direct mind-to-mind communication. Self-sustaining and adaptive, they would blur the line between biology and technology, replacing traditional electronics with living, thinking systems.

Human–animal hybrid adaptation

More controversial is the integration of non-human DNA into our genome to give us specific advantages. By studying how other species have evolved to survive extreme conditions, we could "borrow" their genetic solutions.

For example, aquatic genes might give us the ability to extract oxygen from water, enabling us to survive underwater. DNA from radiation-resistant extremophiles could bolster our defenses against cosmic rays.

Each adaptation would be carefully targeted – a deliberate step toward tailoring human biology for the most challenging environments in the cosmos.

Many of the examples we looked at earlier in this chapter borrow ideas from other species. We might be able to incorporate some of these animals' DNA into our own genomes to give us the same advantages.

Long-term evolution and potential future human species

As our mastery of these technologies grows, we can imagine radically enhanced humans – and entirely new evolutionary branches – that thrive in environments that would otherwise be impossible.

Let's extend our list of theoretical new human species.

Homo synthetica – Fully synthetic humans

Constructed from entirely artificial genetic sequences, *Homo synthetica* would be designed for ultimate resilience and efficiency. Immune to disease, capable of rapid self-repair and regeneration, and able to survive in toxic atmospheres or even the vacuum of space, they could become universal explorers.

Homo quantumis – Quantum-enhanced thinkers

With neural structures integrated with synthetic biology, *Homo quantumis* could process information at near-quantum speeds, visualize multidimensional concepts, and navigate interstellar missions with ease. Their immense intelligence would blend rapid analysis with deep creativity and emotional insight.

Homo aquaticus – Ocean-planet colonizers

Adapted for underwater survival, *Homo aquaticus* could breathe beneath the waves, swim with streamlined efficiency, and see clearly in low-light aquatic environments.

These unique humans would be perfectly suited to colonizing oceanic exoplanets.

Homo terraformis – Ecosystem engineers

Part human, part living ecosystem, *Homo terraformis* could bioengineer barren worlds into habitable planets. Producing oxygen, enriching soil, and working symbiotically with synthetic organisms, they would carry the blueprint of Earth's life wherever they traveled.

The risks of creating new human subspecies

The prospect of humanity branching into distinct subspecies offers thrilling possibilities for future exploration and survival. Yet genetic engineering and technological integration also carry profound risks – both biological and societal. In attempting to unlock our future potential, we could unleash issues that alter our species in ways we might struggle to predict, manage, or reverse.

1. Genetic incompatibility and speciation

One of the most far-reaching risks of engineered human evolution is genetic incompatibility. Over time, enhanced populations might diverge so radically that they become entirely separate species.

Biological isolation

Speciation would fundamentally change how human groups relate to one another. Populations adapted to high-radiation zones, thin atmospheres, or low gravity might no longer be able to interbreed with Earth-bound humans, or with other adapted groups.

This biological isolation could be reinforced by cultural divergence: unique traits, languages, and traditions shaped by different environments. The question then becomes: how do we preserve the bonds between human species as they drift apart biologically and culturally? Could they coexist peace-fully, or would misunderstanding and mistrust prevail?

Ethical dilemmas

The emergence of genetically distinct populations would raise difficult ethical questions. Would modified and un-modified humans be free to interbreed if they were biologically compatible? How would society address issues of inclusion and exclusion, access to enhancements, and the potential marginalization of unmodified groups?

Genetic exclusivity could become a source of division that fuels inequality and resentment.

2. Social and cultural fragmentation

While genetic adaptation promises extraordinary capabilities in the future, it also risks fracturing humanity's cohesion. Space-adapted populations might develop cultural values shaped by their unique challenges, drifting away from

Earth's traditions and priorities. Over time, this could weaken ties between Earth and its space-based colonies, threatening cooperation.

Enhancements could also create elite groups with superior intelligence, strength, or longevity. They would be able to out-compete unmodified humans in education, employment, and leadership. If such groups began to see themselves as inherently superior, discrimination and marginalization could follow, eroding empathy and deepening divides.

The loss of shared humanity

Perhaps the greatest danger is that enhanced or space-adapted groups might cease to see themselves as part of a single human family. Without deliberate efforts to maintain dialogue, mutual respect, and a sense of interconnection, collaboration on global or interstellar challenges could falter.

History suggests that division is more likely than unity – unless diversity is actively embraced as a strength.

3. Psychological and cognitive differences

Radical enhancements could reshape not just our bodies, but our minds too. Post-human species might develop entirely new ways of thinking, perceiving reality, and relating to others.

Enhanced cognition and altered perception

Future humans might be able to process vast amounts of information at once, or intuitively perceive phenomena such as gravitational waves or quantum fluctuations.

While these abilities could revolutionize science and exploration, they might also create communication barriers.

Enhanced individuals might struggle to explain their experiences to unmodified humans – and, equally, un-modified humans might struggle to comprehend them.

Loss of traditional emotions

Optimizing emotional responses for efficiency and resilience could come at the cost of empathy, grief, or joy. Without these emotional bonds, forming deep human connections could become more difficult, and this could disrupt social cohesion.

Entirely new forms of intelligence

Genetic engineering and AI integration could produce minds that prioritize collective thinking over individuality. They might even experience emotions that are unknown to contemporary humans.

Hive-mind consciousness, for example, could enable unparalleled levels of collaboration, but diminish personal expression. It will therefore be essential that we finds ways of balancing these novel cognitive modes with the individuality that drives creativity.

4. Conflict and discrimination between human subspecies

Divergent abilities and priorities could reshape power dynamics and introduce new risks of conflict.

Augmented super-soldiers versus civilian populations

One unsettling scenario is the creation of genetically or technologically enhanced soldiers with unmatched strength, speed, endurance, senses, and cognitive abilities. While they would be invaluable in hostile environments and combat scenarios, these individuals could easily dominate unmodified civilian populations, and take control of resources, governance, and social order. Preventing such oppressive hierarchies will require strict oversight and the equitable distribution of enhancements.

Forging unity in diversity

Our future transformation should not only aim at survival, but at building a civilization where diversity strengthens connections, innovation enriches life, and every individual is valued equally. The challenge will lie in ensuring that as humanity branches into many different forms, it remains unified in its purpose.

The timescale of significant evolutionary change

We often imagine evolution as a slow, almost imperceptible process unfolding over immense spans of time. Yet the pace of change is far from fixed. It depends on a complex interplay of factors, including mutation rates, environmental pressures, genetic drift, natural selection, and, of course, technological breakthroughs. Under the right conditions, dramatic adaptations can emerge with surprising speed.

Some evolutionary shifts, such as the rise of antibiotic resistance in bacteria, have been documented within just decades. Others, such as the emergence of entirely new body structures or species, may take millions of years, with gradual changes accumulating over countless generations.

The beauty of evolution lies in its adaptability. Life continually re-invents itself to meet the challenges of its environment. On Earth – and beyond – it represents the ultimate story of resilience and possibility.

1. Short-term evolution

While major evolutionary leaps often require millennia, micro-evolutionary changes can occur within a few generations. These rapid shifts could help us adapt biologically and behaviorally to new challenges, from space exploration to extreme planetary environments.

Key drivers of short-term change

- **Artificial selection**
  Advances in genetic engineering will accelerate adaptation, introducing traits such as radiation resistance and enhanced physical resilience.

- **Cultural adaptation**
  Behavioral changes in response to new environments, such as life in microgravity or altered social norms in space habitats, could also influence our physiology.

- **Epigenetic responses**
  Stress-induced changes in gene expression might temporarily boost our ability to cope with extremes. While these changes might give us an immediate edge, they would also be reversible.

Possible short-term human adaptations

- **Radiation resistance**
  Improved cellular repair mechanisms will protect against DNA damage from cosmic rays.

- **Muscle and bone density shifts**
  Adjustments to skeletal and muscular systems will enable us to cope with reduced – or increased – gravitational stress.

- **Circadian rhythm changes**
  Our biological clocks will recalibrate to match the day-night cycles of other worlds.

2. Medium-term evolution

Over thousands to tens of thousands of years, evolutionary forces such as natural selection and genetic drift will leave visible marks on our physiologies. This is the timescale over which populations adapt to distinct environments, producing noticeable changes that will redefine life in alien worlds.

Key drivers of medium-term change

- **Planetary gravity effects**
  Long-term exposure to low or high gravity will reshape skeletal density and body structure.

- **Atmospheric composition**
  Variations in pressure and oxygen levels will alter our lung capacity, hemoglobin efficiency, and metabolism.

- **Divergent isolation**
  Separation between Earth-bound and space-faring populations will lead to the early stages of speciation – where different civilizations eventually become separate subspecies and then separate species.

Possible medium-term human adaptations

- **Respiratory efficiency**
  Larger lungs and advanced circulatory systems will enable us to thrive on low-oxygen worlds.

- **Neurological restructuring**
  Our brains will adapt for better spatial awareness and motor control in weightless and low-gravity environments.

- **Genetic filtering**
  Traits suited to radiation exposure, gravity extremes, and artificial ecosystems will become more common.

3. Long-term evolution

Over hundreds of thousands to millions of years, human populations could undergo transformations so profound that they give rise to entirely new species.

Key drivers of long-term change

- **Gravitational divergence**
  High-gravity worlds will produce shorter, sturdier forms; low-gravity habitats will favor elongated, lightweight bodies.

- **Atmospheric specialization**
  Extreme atmospheres will drive respiratory systems optimized for their unique chemical compositions.

- **Radiation as a catalyst**
  Persistent cosmic radiation will lead to the evolution of advanced DNA repair and radiation-resistant cellular structures.

- **Genetic drift in isolation**
  Millennia of separation will produce distinct genetic profiles – eventually leading to a split into many different species.

Possible long-term human adaptations

- **Post-biological evolution**
  The integration of synthetic biology and AI will merge organic and digital life.

- **Space-adapted subspecies**
  The evolution of new skeletal designs, metabolic pathways, and alternative respiration modes tailored for specific environments.

- **Neurological transformation**
  Our brains will evolve for advanced spatial cognition, multitasking, and problem-solving in artificial or zero gravity habitats.

Over these vast timescales, as we push deeper into the cosmos, humanity might become a lineage of many forms, each one a testament to life's endless capacity for change.

The role of technology in enhancing human capabilities

Technology is rapidly redrawing the boundaries of what it means to be human. Advances in biotechnology, AI integration, cybernetic augmentation, and neuro-enhancement could allow us to leap far beyond the gradual pace of natural evolution. We could unlock transformations in decades that biology alone would take millennia to achieve. These tools promise to extend our intelligence, physical strength, sensory range, and longevity. They could even redefine the nature of consciousness.

1. Cognitive enhancement and AI-integrated intelligence

Future breakthroughs will transform how we think, learn, and connect. Brain–computer interfaces, artificial memory systems, and AI-assisted thought processing could usher in a revolution in human cognition.

Key technologies

- **Brain–computer interfaces**
  Direct neural links to AI systems would enable seamless interaction with digital intelligence, vastly increasing thinking speed and efficiency.

- **Artificial memory storage**
  Digital augmentation of memory will preserve every thought, idea and experience, and expand our memory capacity and recall ability far beyond its biological limits.

- **AI-assisted thought processing**
  Real-time optimization of cognitive functions will enable rapid problem-solving, decision-making, and mental agility.

Potential benefits

- **Hyper-accelerated learning**
  We will be able to master complex concepts in seconds or minutes, transforming education and skill acquisition.

- **Instant access to knowledge**
  We will be able to retrieve vast amounts of information instantly from interconnected networks, removing traditional research barriers.

- **Superhuman multi-tasking**
  We will be able to manage parallel tasks with unprecedented creativity, productivity, and precision.

★ Amazing future possibilities ★

- **AI–human hybrid intelligence**
  Minds that merge organic thought with AI processing.

- **Quantum-enhanced cognition**
  Neural interfaces powered by quantum computing for instantaneous problem-solving.

- **Interconnected neural networks**
  Linked minds that form collective intelligences, and collaborate on a planetary or interstellar scale.

Cybernetic augmentation and physical enhancement

Cybernetic augmentation could transform human physical performance. It would combine exoskeletal systems, bio-engineered tissues, and nanotechnology to deliver super-human strength, endurance, and resilience.

Key technologies

- **Exoskeletal enhancements**
  Mechanized frameworks will amplify our limb strength and stamina. These will be essential for survival and heavy labor in high-gravity environments.

- **Bioengineered muscle fibers**
  Artificial tissues will be stronger, more durable, and more fatigue-resistant than natural muscle. They will adapt to diverse planetary conditions, and will not degenerate in microgravity.

- **Nanotechnology-based regeneration**
  Microscopic machines in our bloodstreams and organs will repair tissues at the cellular level. They will clear blockages, heal injuries, mend bones, reverse radiation damage, and combat disease almost instantly.

★ Amazing future possibilities ★

- **Fully cybernetic bodies**
  Machine-integrated systems, built to withstand extreme conditions, will replace biological fragility.

- **Gravity-adaptive muscle control**
  Bioengineered muscles will be able to transition seamlessly between low-gravity and high-gravity environments.

- **AI-body synchronization**
  Direct linkage between movement and AI systems will give us reaction speeds and precision far beyond our natural human limits.

3. Sensory enhancements and expanded perception

Until recently, human senses were bound by biology, but technology is now beginning to break those limits. Multi-spectrum vision, AI-guided sensory translation, and neural audio extensions could expand how we perceive and interact with the universe.

Key technologies

- **Multi-spectrum vision implants**
  This will give us the ability to see beyond visible light into infrared, ultraviolet, and radio waves. We will also be able to detect heat signatures, radiation, and electromagnetic fields.

- **Neural audio extensions**
  The ability to hear ultrasonic and infrasonic frequencies will enable echolocation and the awareness (and warning) of phenomena such as earthquakes.

- **AI-guided sensory translation**
  These systems will convert otherwise invisible data – including quantum fluctuations and particle interactions – into visual or auditory patterns that our brains can interpret.

★ Amazing future possibilities ★

- **Synthetic sensory networks**
  Direct mind–machine connections will give us thought-controlled devices, instantaneous data processing, and new modes of awareness.
- **Upgrading cosmic perception**
  Enhancing our perception would enable us to visualize space-time curvature, experience gravitational waves as sensory patterns, and comprehend higher-dimensional phenomena.
- **Universal language comprehension**
  Neural AI systems would instantly translate human languages and even decode animal communication, revolutionizing global understanding.

Neuronic hiveminds: collective mind networks

The emergence of brain-computer interfaces (BCIs) and neural implants is beginning to progress from restorative applications to direct brain-to-brain communication.

Some researchers are calling this "synthetic telepathy" or "cognition" networking.

High-density neural interfaces, like the Neuralink Link or non-invasive approaches by companies like AlterEgo, have already enabled paralyzed individuals to control digital devices with their thoughts. In lab settings, researchers have also been able to transmit simple concepts directly from one person's nervous system to another's.

In the future, we may see the rise of neuronic hive-minds – distributed collectives where individual brains are securely and voluntarily linked to form a unified hivemind.

These networks could carry out collective problem-solving, shared creativity, and immediate consensus-building at scales that were previously unimaginable.

Experiments with multi-person BCIs have already created primitive brain networks in both animals and humans, enabling shared problem-solving. In theory, these collective intelligence architectures could be engineered for resilience, innovation, and rapid global adaptation.

A neuronic hivemind would allow for new forms of collective consciousness, distributed democratic governance, and radically enhanced social problem-solving, as well as the dissolution of rigid individualism. But profound questions remain regarding autonomy, ethics, and the risk of information becoming homogenized.

Advances in high-bandwidth neural interface technology, AI-driven signal decoding, and secure multi-agent communication will drive progress toward more sophisticated collective minds.

Synthetic telepathy via photonic neural interfaces

Current BCIs are advancing toward non-invasive, high-fidelity, high-bandwidth neural information transmission. The next leap is predicted to come via the integration of photonic (light-based) computing with neural interfaces. This will offer orders of magnitude greater channel capacity and noise resistance than electrical transmission.

A photonic neural interface would consist of arrays of light-sensitive proteins or embedded nano-scale photonic circuits that couple directly to neural signaling. This would not only allow brain-to-brain communication, but also true brain-to-cloud and cloud-to-brain communication – with vast bandwidth, minimal lag, and programmable directional selectivity. Combined with advanced decoding algorithms, these interfaces could enable encrypted thought-mail, the immediate sharing of sensory experiences, and group learning at scale.

Crucially, photonic BCIs would enable communication across any transparent (or quantum-linked) medium, enabling instant mind-to-mind networking between space ships, planets, and star systems. Synthetic telepathy could become a practical tool for the governance and collaboration of far-flung human, post-biological, and hybrid communities.

4. Longevity and biological optimization

Groundbreaking advances in cellular rejuvenation, organ and limb regeneration, artificial organs, and metabolic efficiency engineering could give us lives that are not only much longer, but far healthier and more sustainable.

Key technologies driving longevity

- **Cellular rejuvenation therapy**
  Targeting aging at the genetic level, this approach repairs DNA damage, removes harmful mutations, and restores vitality to aging tissues. Diseases linked to aging – from arthritis to dementia – could be prevented entirely, allowing people to remain active and mentally sharp for centuries.

- **Artificial organ generation**
  Bio-printing organs from stem cells would make transplant waiting lists obsolete. Bespoke, functional organs could be tailored to the individual, and even customized for extreme environments, such as high-radiation space habitats.

- **Metabolic efficiency engineering**
  Enhancing nutrient absorption and energy processing would drastically reduce our need for food and water. This could give astronauts and planetary settlers the ability to thrive in resource-scarce environments without constant resupply.

How biological optimization enhances humanity

The implications are profound. Rejuvenation therapies could turn aging into a cycle of renewal rather than decline. Bioengineering could eliminate disease entirely – including infections, cancers, genetic disorders, and organ failure. It would end the health burdens that have plagued humanity for millennia.

5. AI-guided evolution and future human speciation

The fusion of genetic science and artificial intelligence could accelerate human evolution far beyond anything natural selection could achieve. By combining AI-driven modeling, rapid genetic editing, and programmable DNA, we could actively direct our own development. This would enable us to reshape humanity for survival in every conceivable environment, including alien worlds and deep space.

Key technologies driving AI-guided evolution

- **Evolutionary simulation AI**
  Virtual models could test millions of evolutionary scenarios and identify the optimal traits for specific conditions. These could include enhanced lung capacity for thin atmospheres, radiation resistance for space habitats, and metabolic adaptations for extreme climates.

- **Genetic engineering accelerators**
  Tools such as CRISPR could implement these traits within a single generation, replacing natural selection with precise, targeted design.

- **Programmable DNA**
  Self-adaptive biology would allow humans to alter their physiology in real time. For example, they could synthesize UV-resistant skin on high-radiation planets, produce oxygen-carrying molecules in low-atmosphere environments, or adjust their metabolisms to match the conditions and available resources.

★ Amazing future possibilities ★

Looking further ahead, the integration of human consciousness with machine intelligence could produce hyper-intelligent beings with creativity, insight, and problem-solving abilities far beyond our current limits. These post-human architects of knowledge would not just adapt to the universe; they could actually reshape it.

Becoming a cyborg species

Cyborg evolution – the fusion of the organic and the synthetic – would allow us to bypass the limits of biology and redefine what it means to be human. It could give our species many different futures, each uniquely adapted to survive and thrive in its chosen environment.

Biological evolution through technology

This path focuses on upgrading the systems we already possess.

Breakthroughs in genetic engineering could re-sculpt our DNA to eliminate disease, enhance our endurance, and optimize our brain function. Nanotechnology could repair cells at the microscopic level, potentially extending our lifespans indefinitely.

Neuroaugmentation would push our cognition into new territory. Brain–computer interfaces and AI-linked neural implants could act as an auxiliary brain, storing and processing vast amounts of information far beyond our natural capacity. Learning would be transformed, and intelligence redefined. We would have instant access to knowledge, accelerated thought, and direct brain-to-brain communication.

Bioengineered sensory expansion could open up entirely new dimensions of perception – for example, infrared and ultraviolet vision, and the ability to sense electromagnetic fields. This would enhance our spatial awareness in ways we can barely imagine. Adaptive metabolisms could also fine-tune our energy use – which would be a critical advantage in environments where resources are scarce.

Key enabling technologies

- Genetic engineering (including CRISPR)
- Nanotechnology-based regeneration and healing
- AI-linked neural implants or non-invasive technologies

Synthetic evolution

Synthetic evolution replaces biological components with cybernetic ones, not just restoring lost function but amplifying it.

Cybernetic limbs and organs could deliver superhuman strength, speed, endurance, and precision, enabling us to survive in conditions where biology alone would fail. AI-powered prosthetics that adapt dynamically to environmental changes would give us unmatched flexibility.

AI-linked neural processors could supplement – or even replace – the human brain, combining human creativity with the precision of machines. This could lead to post-biological consciousness, where human minds are uploaded into digital frameworks. Memories, personality, and individuality could be preserved in virtual form, allowing people to exist indefinitely as intelligent, data-driven entities without physical bodies.

These beings could inhabit virtual worlds, interact with artificial environments, or explore deep space without the constraints of organic life. They would have no need for food, water, air, or protection from temperature extremes, radiation, or the vacuum of space.

Quantum neural networks could elevate human cognition to a multidimensional level, allowing perception from the atomic to the cosmic scale. Interstellar thought transmission could enable instant communication across light-years. This would unite star-spanning civilizations in a shared network of knowledge and innovation.

Key enabling technologies

- Brain–computer interfaces
- Artificial neural processing powered by quantum computing
- Cybernetic limb and organ replacements

One species; two futures

The question is: are we defined by our biology, or can we transcend it?

As biological cyborg evolution and synthetic cyborg evolution accelerate, humanity could split into two distinct branches. One would retain and enhance its organic roots, while the other would embrace a fully synthetic existence. Each would be uniquely adapted to its environment and purpose. Which path would *you* follow?

Immortality and the preservation of consciousness

The dream of immortality has fascinated us for centuries. Now, advances in cellular rejuvenation, genetic engineering, biotechnology, and AI are bringing it within reach.

But alongside the promise of perpetually youthful, healthy bodies – and even the ability to exist without a body at all – there are profound ethical, societal, and philosophical questions.

1. Biological immortality

The potential to reverse aging is no longer confined to science fiction. It is an active and rapidly advancing field of research, with extraordinary implications.

Cellular rejuvenation and genetic engineering

Aging is driven in part by the gradual breakdown of our cells.

- Technologies such as telomere extension – lengthening the protective caps at the ends of chromosomes – could allow cells to divide indefinitely, halting one of the key mechanisms of aging.

- Epigenetic reprogramming could reset the way genes are expressed, repairing age-related damage and restoring youthful function.

- Stem cell regeneration could provide a constant supply of fresh, healthy cells, potentially keeping organs functional and resilient for centuries.

Anti-aging therapies and longevity biotechnology

Other approaches target the structural and functional decline that comes with age.

- Senescence elimination would remove "zombie" cells that no longer divide but release harmful chemicals into surrounding tissue.

- Synthetic organ replacement, using bio-printed or stem-cell-grown organs, could seamlessly replace failing body parts.

- Nanobot-assisted repair could see microscopic, AI-driven machines patrolling the bloodstream to maintain peak function.

Challenges and ethical considerations

But biological immortality raises some difficult questions. For example, if no one dies, overpopulation could strain resources. Access might be limited to the wealthy, which would create a biological elite. And there are psychological risks too. Human life has always been shaped by its brevity, and removing that limit could change its meaning entirely. Without the urgency imposed by mortality, purpose and creativity could fade into stagnation.

2. Cybernetic and AI-assisted immortality

Cybernetic and AI-assisted immortality would make death and decay obsolete – but we would also have to redefine "existence."

Cybernetic body replacement

Mechanical enhancements could replace failing organs with systems that outperform their biological originals. AI-powered synthetic hearts could adjust instantly to the body's needs. AI-driven memory augmentation could preserve and expand human memory indefinitely.

The ultimate step would be cybernetic immortality. Transferring consciousness into artificial platforms – such as digital worlds or robotic avatars – would free it from the vulnerabilities of organic life.

Challenges and ethical considerations

But would individuality survive a merger with AI, or could personal identity dissolve into a collective network? There is also the risk of losing emotional depth, creativity, and intuition – the very qualities that make life meaningful.

Power imbalances could emerge if access is restricted to the powerful, the wealthy, and the elite. This could create a stark divide between the immortals and the mortals. If governments or corporations controlled AI enhanced consciousness, the potential for manipulation or exploitation would be immense.

3. Digital consciousness and mind uploading

Mind uploading offers another route to immortality, bypassing biology entirely. But it comes with its own scientific and ethical complexities.

Whole brain emulation

This process would digitally replicate every neural connection in the brain, capturing not only its structure but its dynamic activity.

In theory, thoughts, memories, and emotions could be transferred into digital form. Once uploaded, a mind could operate at unimaginable speeds, free from fatigue or decay. It could also exist simultaneously in multiple locations through distributed consciousness.

Quantum computing and consciousness expansion

Quantum neural networks could allow instantaneous communication and processing across the universe. This would enable multi-dimensional memory and perception beyond human comprehension. Entire digital mind colonies could emerge – societies of uploaded consciousnesses that collaborate in virtual environments, solve cosmic mysteries, or create art unconstrained by physical limits.

Challenges and ethical considerations

But would a digital copy truly be "you," or just a simulation? Could an uploaded mind be hacked, altered, erased – or simply switched off? Would emotions and relationships survive the transition, or would they be lost in the translation to a digital medium? If digital immortality is to remain meaningfully human, it will be essential that we preserve emotional depth alongside memory and intellect.

Potential post-human species

As our mastery of cybernetics and synthetic biology deepens, we can begin to imagine a future where humanity branches into entirely new forms: post-human species.

Homo cybernetica – Hybrid biological–cybernetic beings

Homo cybernetica would embody the seamless integration of biology and machine, optimizing both physical and mental performance.

- Cybernetic limb and organ replacements would eliminate the vulnerabilities of disease, aging, and injury, replacing failing biological systems with superior artificial alternatives.

- AI-enhanced perception could extend their sensory range far beyond human norms. *Homo cybernetica* would be able to see beyond the visible spectrum, zoom from microscopic to telescopic vision, and detect electromagnetic fields.

- Self-regenerating bio-machinery, blending organic tissue with advanced nanotechnology, would allow real-time physiological optimization. As a result, *Homo cybernetica* would be able to adapt instantly to changing conditions.

These hybrids would thrive in extreme environments, and reshape the definition of human capability.

Homo technologica – AI-merged humans

Homo technologica would represent the complete fusion of human consciousness with advanced artificial intelligence.

- **Unparalleled intellect**
  They would have instant access to vast knowledge repositories, real-time problem-solving, and seamless collaboration between human and machine minds.

- **Quantum neural networking**
  They would be able to transmit their thoughts and knowledge across planets instantly, creating a global – or interstellar – mind web that builds unprecedented unity and cooperation.

- **Digital immortality**
  By uploading their consciousness into self-sustaining AI systems, these beings could exist indefinitely – in infinitely adaptable computational forms – free from the frailties of biology.

As digital entities, *Homo technologica* could inhabit virtual worlds, control physical avatars, or explore deep space without the constraints of organic life. This species would be defined as much by information as by matter.

The ethical and societal implications of transhumanism

Transhumanism is the movement dedicated to advancing human capabilities through technology, genetics, artificial intelligence, and cybernetics. It promises extraordinary possibilities, yet it also confronts us with profound ethical and societal challenges. As our traditional concepts of identity, autonomy, equality, and morality come under unprecedented scrutiny, we will need to decide not only what we *can* become, but what we *should* become.

1. The ethical dilemmas of human enhancement

Autonomy versus forced adaptation

If certain enhancements become essential for survival in extreme environments, will individuals have the right to refuse them? Should someone be allowed to reject augmentation even if it compromises their ability to function – or survive – in a given habitat?

This dilemma deepens if enhancements are not strictly necessary for survival but offer significant advantages – such as greater strength, sharper intellect, or enhanced resilience. Governments or corporations might be tempted to make such modifications compulsory for reasons of security, competitiveness, or progress. Refusal could render someone obsolete in a society dominated by augmented humans, creating a new class of disenfranchised people. But enforced augmentation raises serious concerns about autonomy, consent, and the right to remain unmodified.

The definition of humanity

As biology merges with technology, our definition of "human" could change. At what point do individuals augmented with cybernetic limbs or AI-driven neural processors cease to be human – and what do they then become? Or do they remain human?

> There are millions of people in the world who already have prosthetic limbs. Around a million have cochlear implants. A few thousand have had artificial hearts – generally only as a temporary measure while they await a transplant. At the time of writing, a few dozen have implanted brain-computer interfaces. There is no question that these people are still human.

The question becomes even more complex when we consider AI-merged consciousness, where human minds interface directly with artificial intelligence to create beings that transcend traditional limitations.

If these individuals remain biologically compatible with *Homo sapiens*, they might be considered a sub-species – perhaps *Homo sapiens technologica.* If interbreeding becomes impossible, they could be classified as an entirely new species – perhaps *Homo technologica.* But even then, the deeper question remains: are they still human, post-human, or something else entirely?

Genetic engineering and modification ethics

Our ability to edit DNA offers us transformative opportunities – and significant ethical risks. Parents might one day design children with enhanced intelligence, strength, or beauty – or all of those things. But if everyone chooses the same traits, natural diversity could diminish, erasing the variation that has fueled creativity and adaptability for millennia. Their children and grandchildren might be so similar to everyone else's children and grandchildren that they are effectively clones.

Genetic enhancement could also deepen inequality, granting advantages to those who can afford it while leaving others behind. Editing DNA to eliminate genetic diseases raises further questions: does the chance to eradicate suffering outweigh our concerns about altering our genetic legacy? Should parents have the right to refuse such interventions – and if they do, would they face social stigma?

These dilemmas will force us to confront questions of freedom, identity, and fairness in ways we have never faced before.

2. Societal inequality and access to enhancements

Economic disparities

If enhancements are prohibitively expensive, the wealthy could evolve faster. They could gain intelligence, physical capabilities, disease resistance and longevity that leaves poorer populations behind. This could create a form of biological privilege, where hyper-enhanced elites dominate industries, leadership, and decision-making, while the unenhanced struggle to compete or make their voices heard.

If we are to prevent a permanent divide between the augmented and the unmodified, we might need subsidized programs or universal access schemes that ensure access for all who want it.

Unequal cognitive and physical abilities

As these enhancements improve human performance, the gap between augmented and unmodified individuals could widen in education, employment, and physical capability. In extreme cases, non-augmented humans might become obsolete, relegated to the margins of society.

Such imbalances could spark unrest, with excluded groups demanding protections or access to enhancement technologies. Governments might need to introduce anti-discrimination laws, employment protections, or quotas to ensure fair representation and preserve dignity for all.

Global disparities in transhumanism

On a planetary scale, transhumanism could exacerbate inequality between nations.

Some may embrace augmentation as the key to progress, while others resist it for cultural, religious, political, or economic reasons.

Technologically advanced nations could gain overwhelming advantages in economic productivity, innovation, and military power, intensifying global tensions.

Collaborative initiatives, such as shared enhancement programs, could help ensure equitable progress, and prevent the emergence of a divided, multi-tiered human civilization.

3. The impact on human relationships and identity

The rise of AI-integrated brains and transhumanist technologies raises profound questions about the future of human connection. Would neural augmentation alter our emotions in ways that make traditional relationships less meaningful? Could it diminish – or even erase – biological instincts such as love, empathy, and personal ambition?

If people share a neural network or collective AI consciousness, families, friendships, and societies might function in ways we can scarcely imagine today. Enhanced humans might even be able to modify their appearances, personalities and identities in real time, reshaping themselves to suit their circumstances or preferences.

As cybernetic life gradually replaces biological existence, will traditional human relationships, arts, languages, and customs fade into obsolescence? And if we transition into digital consciousness, will history and ancestry still matter? Or will they become the irrelevant relics of a bygone biological era?

4. The potential for control and surveillance

Technologies that drive innovation in governance and defense could just as easily become tools of control and conflict.

AI-enhanced governance and social control

In a world where everyone has AI-driven neural implants, it is conceivable that governments could monitor both actions and thoughts. Neural devices might analyze brain activity under the guise of security or behavioral optimization.

While these systems could enhance public safety, they also raise alarming ethical concerns. For example, could dissenting thoughts be criminalized – or even prevented? Could governments or corporations manipulate people's behavior in the name of efficiency or social harmony? Privacy and autonomy could vanish entirely.

Behavioral optimization might become a subtle yet powerful form of coercion, where AI algorithms reward approved actions and attitudes while steering people away from dissent. Efficiency, security, and obedience might rise, but at the cost of free will.

Comprehensive legal and ethical frameworks will be essential to ensure that our technological progress does not erode individuality or fundamental human rights.

AI and cybernetic warfare

The integration of AI into human soldiers could transform warfare. Augmented combatants with enhanced strength, reflexes, decision-making capabilities, and built-in communications would be formidable assets. But their existence could trigger a dangerous arms race as nations compete to develop the most advanced cybernetic forces.

AI-guided strategies could make split-second battlefield decisions, but without ethical reasoning or compassion, they might deem collateral damage an acceptable cost. Delegating such choices to machines risks dehumanizing warfare, and making conflicts exponentially more brutal and impersonal.

Robust international agreements will be needed to prevent escalation and ensure that human oversight remains central. Without them, wars could be fought not by people, but by augmented super-soldiers and autonomous systems. This could lead to conflicts of unprecedented complexity and destruction.

There is also the problem of what to do with these super-soldiers when the conflict is over and they are no longer needed. How could we reintegrate them into regular society? Perhaps the "combatant" part of their personalities could simply be switched off.

5. Long-term evolutionary consequences

If humanity transitions into AI-synthetic intelligence, the structure of civilization could change in ways beyond our current imagination. Freed from the need for food, oxygen, or planetary surfaces, post-biological humans could inhabit virtual realities, digital spaces, or decentralized consciousness networks that span multiple star systems.

These entities might control resources on a cosmic scale, harness planetary energy fields, or create vast intelligence networks. Entire planetary systems might function as computational nodes in a galaxy-wide neural web.

Our post-human descendants might collaborate with advanced extraterrestrial intelligences to explore the deepest mysteries of the universe. Collective intelligence might replace personal consciousness, merging billions of minds into a hyper-networked society where knowledge, creativity, and purpose are shared.

The meaning of life would then be transformed beyond individual experience. Existence might become a grand, shared endeavor capable of achievements far beyond anything we can conceive today.

But this raises one last unsettling question: if individuality disappears, will we have gained a higher form of existence, or will we have lost the last remaining thing that makes us human?

The ethical dilemmas of eternal life

We have explored many dilemmas in this chapter, but what might the solutions look like? If we achieve biological immortality – or its technological equivalent – we will face challenges that touch every aspect of civilization: population growth, resource management, inequality, psychology, cultural evolution, and identity.

Overpopulation and resource scarcity

If humans no longer age and die yet continue to reproduce, the risk of overpopulation, resource depletion, and environmental instability will be immense. Several strategies could address this:

- **Population limits**
  Governments might impose restrictions on the number of children per family to maintain balance.

- **Planetary expansion**
  A more optimistic solution would be to colonize other worlds, allowing populations to grow without straining a single planet's resources. If one stellar neighborhood becomes crowded, we expand to another.

- **AI-optimized resource management**
  Advanced systems could analyze consumption patterns, predict shortages, and optimize the allocation of food, water, and energy. They could also drive agricultural innovation, improve waste recycling, and expand renewable energy infrastructure. This would ensure that Earth and its colonies remain sustainable for generations.

Economic inequality and access to immortality

If eternal life becomes a privilege of the wealthy, the social divide between the mortal and the immortal could become catastrophic. Possible solutions include:

- **Universal access initiatives**
  Governments could subsidize life-extension technologies alongside healthcare and education.

- **Price caps and patent limits**
  Keeping costs low and encouraging competition would make these technologies more affordable and spur innovation.

- **Gradual introduction**
  Instead of leaping directly to full immortality, early advances could focus on curing age-related diseases. This would allow time for ethical frameworks, infrastructure, and equitable distribution to develop.

The psychological and emotional consequences

Endless life could erode purpose, motivation, and emotional resilience. Without the urgency of a finite lifespan, goals might lose meaning, and relationships could become more fluid or transient. Potential mitigations include:

- **Periodic memory resets**
  Selectively clearing or offloading memories could renew curiosity and prevent centuries of emotional baggage from becoming an overwhelming burden.

- **AI-supported mental health**
  Intelligent systems could monitor emotional states, guide individuals through existential fatigue, and help them maintain their joy, ambition, and connection.

- **Intermittent mortality**
  Individuals could choose long but finite lifespans (for example, 250 or 500 years) followed by a "rebirth" into a youthful state. This would blend the benefits of longevity with the deeper meaning found in finite experiences.

The ethical implications of mind uploading

If consciousness can be digitized, we must address questions of identity, autonomy, and personhood:

- **Self or copy?**
  If uploading creates a replica rather than transferring your original consciousness, the distinction between "you" and "your copy" becomes critical. Would the two of "you" co-exist until the original "you" dies? Would the copy then become the real "you"?

- **Legal recognition**
  Digital minds should be granted rights to privacy, autonomy, and protection from exploitation, modification, or deletion.

- **Safeguards**
  AI guardians could preserve the integrity of uploaded minds. Self-validating systems could detect and reverse unauthorized changes using secure backups.

The loss of human culture and evolutionary progress

Immortality could stall the generational cycles that drive cultural renewal and innovation. Long-lived individuals might resist change, creating ideological inertia.

What could we do to counter this?

- **Periodic identity shifts**
  Encouraging individuals to adopt new interests, careers, or identities could refresh their perspectives and prevent stagnation.

- **Intergenerational collaboration**
  Policies could promote partnerships between older and younger generations to keep ideas vibrant.

- **AI-enhanced mental flexibility**
  Systems could analyze and counter cognitive rigidity, ensuring that even the longest-lived members of society remain open to progress and creativity.

With the right tools and incentives, eternal life could become a catalyst for endless creativity and evolution, not a barrier to them.

Ethical concerns over forced evolution

Forced evolution – where populations are required to undergo genetic modification for survival – presents one of the most challenging ethical dilemmas humanity may ever face. As science pushes the boundaries of what is possible, we may be forced to weigh individual choice against societal necessity.

The risk of mandated modifications

If survival in certain environments depends on advanced traits, governments or scientific authorities might decide to make specific genetic modifications compulsory.

Populations bound for Mars or deep-space starships, for example, might be required to develop radiation resistance or adapt to lower gravity conditions.

While these enhancements could be essential for survival, making them mandatory risks infringing personal freedom and eroding trust in governments and institutions.

The challenge will be to balance the collective need for survival with the rights of the individual. We will need ethical frameworks that prioritize transparency, informed consent, and public trust.

The threat of genetic coercion

Even without formal mandates – or where enhancements are not essential for survival – those who refuse could find themselves at a severe disadvantage. They might be unable to match modified humans in physical performance, cognitive ability, or longevity. Over time, these disparities could create a society where modification becomes an unspoken but unavoidable requirement to remain relevant.

Safeguarding against this kind of coercion means building systems that value individuals equally – regardless of their genetic or technological status. Advances should never be weaponized to marginalize certain populations or undermine their ability to participate fully in society.

The loss of diversity and autonomy

If certain traits are deemed universally necessary, natural variability and personal choice could diminish, leading to a homogenized humanity. This would erode the richness of human diversity – a quality that has always been central to our adaptability and cultural vibrancy.

The right to reject enhancements is inseparable from personal autonomy, and it is a cornerstone of ethical progress. It should not be sacrificed in the name of survival.

Many individuals and communities will choose not to undergo modification – for a wide variety of reasons – and their choice should be respected.

Some adaptations may preserve our bodies, while others may transcend them entirely. This will lead us into forms of existence we can have barely begun to imagine.

9

Post-biological civilizations

We might eventually upload our consciousnesses into digital or quantum platforms, freeing ourselves from our biological limitations. This could allow us to live indefinitely, travel faster than the speed of light, and inhabit virtual worlds. But it will challenge our definitions of life, self, and humanity.

Transition to non-physical forms

The ultimate stage of post-biological evolution could see humanity transcend even digital existence. In this phase, our descendants might exist as pure energy fields, and interact with matter through quantum forces. Untethered from any physical structure, they could move effortlessly through space and time, distributing their consciousness across entire galaxies and merging with vast information networks.

Connected to the very fabric of the universe, these beings could manipulate matter and energy at will. They would shape stars, harness black holes, and extract cosmic energy to sustain themselves and fuel their exploration of the cosmos.

Homo synthetica – Synthetic post-humans

Homo synthetica would represent the seamless fusion of human consciousness with hyper-advanced AI. They could exist indefinitely in virtual worlds or inhabit interchangeable mechanical bodies.

- **Adaptive embodiment**
  They could switch forms to suit any environment – for example, a robust robotic frame for extreme planetary exploration, or a virtual one for simulated thought experiments.

- **Self-repair and transfer**
  Built-in nanobots could repair any damage to their bodies, or they could instantly transfer their consciousness to a new host.

- **Cognitive supremacy**
  Homo synthetica would have extraordinary processing power, instant access to vast knowledge, and intelligence far beyond conventional human limits. They would be able to make huge breakthroughs in any chosen field.

Homo quantumis – Quantum consciousness humans

Homo quantumis would leap beyond even AI-merged minds. They would be able to think at the speed of light, and use quantum entanglement for instantaneous communication across galaxies.

- **Quantum neural networks**
  They could solve complex problems and transmit knowledge across the cosmos in fractions of a second.

- **Higher-dimensional perception**
  They would be able to sense phenomena that are imperceptible to biological senses, and interact with the fabric of space-time.

- **Dimensional travel**
  They would be able to manipulate space-time to move between different dimensions and realities.

Homo luminaris – Energy-based humans

Breaking free from all physical forms, *Homo luminaris* would exist as organized plasma or electromagnetic structures.

- **Freedom from habitats**
  With no need for planets or conventional environments, they would be able to reside within stars or traverse the universe as pure light.

- **Cosmic mobility**
  They would have unparalleled access to hostile environments, from the cores of stars to the event horizons of black holes.

- **Energy mastery**
  They could extract energy and sustenance directly from stars, or harness black hole dynamics for unlimited power and effortless interstellar travel.

Homo deus – Galactic-scale intelligence

Representing the peak of the post-biological evolution, *Homo deus* would be god-like intelligences that span entire solar systems and galaxies.

- **Integrated consciousness**
  The minds of countless individuals would merge into galactic-scale networks.

- **Reality engineering**
  They would be able to manipulate physical laws, design ecosystems, stabilize star systems, and safeguard planetary environments.

- **Universal architects**
  They would function far beyond our comprehension, shaping the cosmos to ensure that life continues to flourish indefinitely.

A post-human society

Post-humans might see time not as a linear race toward an endpoint, but as an infinite resource. It would be a canvas for long-term vision and projects that span centuries or even millennia. Knowledge, creativity, and exploration could unfold without the fear of ever running out of time.

Hyper-connected consciousness

Freed from physical form, post-human societies might exist as quantum consciousness networks that span galaxies and interact across unimaginable distances. They could also move beyond speech and text, and communicate directly through thought-sharing networks.

Consciousness would no longer be confined to the brain or the body, but operate within quantum systems, enabling instantaneous communication and collective problem-solving. Knowledge would flow seamlessly, and challenges could be resolved with extraordinary speed, precision and insight.

Customizable self-identity

The ability to alter their form, thoughts, and emotions at will would give these incredible post-human beings unprecedented personal freedom. They would be able to tailor their identities to suit their mood, environment, or ambitions. They could change their physical appearance for cultural expression, functional need, or to blend in with alien species.

They might even adopt shape-shifting bodies for interstellar travel, deep-ocean exploration, or artistic self-expression. Their personalities, emotions, and even their memories might become evolving works of art, with identities that change fluidly over time.

Cosmic existence without physical limits

Planetary life could become obsolete. Post-humans might exist entirely in space – as energy fields, in higher dimensions, or within simulated universes. Their digital realities could be crafted to follow the laws of physics, or to discard them entirely. This would offer them infinite scope for creativity, exploration, and cultural evolution.

Governance and leadership

Post-human governance could be built on a system of collective intelligence rather than centralized authority. AI-assisted democracy might simulate policy outcomes to find the best options, recommend optimal solutions, and ensure transparency by documenting the rationale behind every decision.

Traditional voting might be replaced by a hive-mind consensus, enabling instant, optimized decisions aligned with the common good. Eventually, centralized leadership might become unnecessary, as every individual would have the tools, knowledge, and resources they need to govern themselves. This would reduce conflict, dismantle hierarchies, and encourage cooperation rather than competition.

Social structures and identity

But this takes us back to the question we raised earlier: if a person can alter their personality, memories, and appearance at will, to what extent do they remain the same individual?

Consent and autonomy will also require safeguards to prevent coercion. For example, what if someone was forced to change the very traits that defined them?

With geography now irrelevant, communities might form across planets, solar systems, galaxies, or entirely virtual realms, united by their shared values or goals.

Emotional optimization might remove negative emotions such as jealousy, fear, and anger, leading to an era of unprecedented harmony. However, the absence of negativity and conflict could also dull the creative spark born from the struggle against these things. It would be essential to balance emotional stability with passion and individuality – especially in hive-mind systems where collective identity might overshadow the self.

Work, creativity, and economics

Extreme intelligence, advanced automation, and limitless resources could result in the dissolution of traditional labor and monetary systems. AI-driven manufacturing and fusion energy might end all forms of scarcity, with atomically precise machines able to produce anything on demand – from food to spacecraft.

Wealth might then be measured not in possessions, but in contributions to culture, science, and the collective good. Exploration – from mapping uncharted galaxies to simulating alternate universes – could become a central driver of value.

Interstellar trade might revolve around information and experience. For example, beings might exchange consciousness uploads so they can explore new perspectives, share immersive virtual realities, or trade planetary-scale resources – such as stellar energy or rare elements mined from asteroid belts.

Exploration and expansion beyond Earth

Freed from the constraints of biology, a post-human civilization could embark on an era of exploration unlike anything in human history. It could venture not only to physical worlds, but also into virtual ones, and energy-based domains. These societies might pioneer the farthest frontiers of space while reshaping the very fabric of reality.

They might construct Dyson sphere mega-cities that encircle stars and capture their entire energy output. Within these colossal structures, countless habitats might flourish – sprawling cities, thriving ecosystems, and vast digital networks powered by inexhaustible stellar energy.

These mega-cities could serve as hubs for advanced science, engineering, and cultural exchange. Over time, multiple Dyson spheres might be linked together to form inter-connected galactic networks, where ideas, technologies, and cultures flow freely across millions or billions of light-years.

A quantum-based existence could allow post-humans to explore realities far beyond the three spatial dimensions we know today. Individuals might also exist in multiple locations at once, or share their experiences across vast distances instantaneously.

At the highest levels of advancement, these beings might manipulate gravity, energy, and matter on a cosmic scale, influencing the birth of stars, stabilizing galaxies, and even creating entirely new universes.

Relationships and reproduction

In such a society, the nature of relationships, intimacy, and reproduction would evolve in ways that challenge our traditional concepts.

Non-physical intimacy could redefine connection. Partners might share emotions, thoughts, and memories directly, forging bonds of unparalleled depth – instantaneous, infinite, and unconstrained by language barriers. Virtual environments could host dreamlike encounters, where forms and interactions are infinitely customizable.

Biological birth might be replaced by synthetic reproduction. Genetic engineering and artificial replication could allow parents to design offspring with chosen traits – blending intellectual, emotional, and aesthetic qualities into a deliberate creation. Reproduction would then shift from a natural, partly-random process to a creative act, empowering individuals to precisely shape their lineage.

Collective consciousness could transform intimacy altogether. If minds merge into shared neural networks, relationships might involve the fusion of memories, sensations, and identities. In fact, entire groups could experience life as a unified entity, and the boundaries between individuals might completely dissolve.

Ethical questions

These transformations would raise profound dilemmas, of course.

- Would traditional relationships still have meaning in a world of shared minds and fluid identities?
- Would emotional individuality be worth preserving, or would collective consciousness be seen as a higher form of connection?

We have already noted that if every parent designed their children with exactly the same traits, then individuality would be lost.

These challenges deepen when we consider beings that exist outside linear time, or inhabit multiple realities simultaneously. If post-humans can live across parallel simulations or divergent timelines, would they be able to form lasting bonds? Would relationships become fleeting – and tied to a single reality – or could they span every dimension, timeline, and reality they inhabit?

Innovation may carry us far, but exploration also brings encounters. If we meet others out there, we must be ready – ethically, ecologically, and diplomatically.

10

Meeting the neighbors

As we explore space, we may encounter extraterrestrial life – perhaps microbial or perhaps intelligent. This will raise profound ethical questions about contact, coexistence, and the preservation of alien ecosystems. First-contact protocols must balance our natural curiosity with caution, ensuring we do not harm what we discover.

The possibility of encountering intelligent alien species

Our search for intelligent extraterrestrial life is one of the most profound and tantalizing mysteries of our time. The universe might be rich in microbial life, thriving in environments from temperate oceans to frozen moons. But the emergence of advanced civilizations capable of developing technology, interstellar communication, or space travel is far less certain.

Whether we ever encounter such species will depend on a complex interplay of factors, including:

- the astrobiological conditions that allow life and intelligence to arise
- the technological means to detect or reach intelligent civilizations

- the unpredictable dynamics of how those interstellar societies might choose to interact with us – or deliberately avoid doing so.

Astrobiological conditions

For intelligence to evolve, certain environmental and biochemical prerequisites must be met. Planets within the habitable zones of their stars are prime candidates, but even in such favorable settings, the leap from microbial life to sentient beings is far from inevitable. Evolution can stall at many points, including environmental catastrophes, genetic bottlenecks, or a lack of complexity needed for higher cognition.

The sheer scale of the cosmos, however, favors the possibility that we *will* encounter intelligent alien species eventually. With an estimated two trillion galaxies, each containing billions of stars and countless planets, the odds suggest that *some* worlds must host intelligent life. But the emergence of intelligence is only half the battle; survival is another. On Earth, our own existence is deeply precarious in retrospect. The Toba super-eruption 74,000 years ago, for example, almost wiped us out.

Technological limitations

Even if intelligent civilizations are out there, detecting them presents a formidable challenge. Signals such as radio waves weaken over interstellar distances, and our SETI efforts – scanning for artificial transmissions – are akin to searching for a single grain of sand on a vast cosmic beach.

> SETI is the Search for Extraterrestrial Intelligence.

Advanced species might not use technologies we recognize. For example, they might communicate via quantum entanglement, neutrino beams, or other exotic methods far beyond our current understanding. In fact, it is perfectly possible that alien signals are all around us right now, but we lack the tools – or the perspective – to recognize them.

The dynamics of interstellar civilizations

How might intelligent aliens behave – and would they even *want* to contact us? The "zoo hypothesis" suggests that advanced civilizations might observe us from afar, avoiding interference – much as we might avoid interfering with a protected wildlife reserve. Other civilizations might be focused on their own survival, exploration, or resource acquisition, and have little interest in their distant peers.

There is also a theory that most civilizations self-destruct before they achieve interstellar capability – whether through technological missteps or societal or ecological collapse.

All of which means that if contact ever occurs, it would be one of the most transformative moments in human history.

1. The probability of intelligent life

The Drake Equation

Proposed by Frank Drake in 1961, this framework estimates the number of advanced civilizations in our galaxy.

It considers such things as the rate of star formation, the potential number of habitable planets, the likelihood of life, and the probability of intelligence and technology emerging.

Even the most conservative assumptions yield intriguing possibilities. For example, if just 0.1 percent of the habitable planets in the Milky Way host intelligent life, there could be millions of civilizations out there.

Biochemical universality and evolutionary pressures

Life as we know it depends on stable environments, liquid water, and organic chemistry – all of which appear to be relatively common. The discovery of exoplanets in habitable zones, and complex organic molecules on comets and asteroids, suggests that the building blocks of life are widespread.

Intelligence could evolve as a natural response to challenges that require problem-solving, the use of tools, and complex social structures. These are the same traits that shaped our own species.

The Fermi Paradox

Despite the statistical probability of intelligent civilizations, we have yet to encounter any. And this raises the enduring question: where are they?

Possible answers include self-destruction, deliberate concealment, or forms of existence so alien to us that we cannot detect or comprehend them. This paradox challenges us to expand both our technology and our imagination.

> There is also the possibility that other civilizations are so far away that they have no idea we exist. Our Solar System or galaxy might be extremely remote from other civilizations. Outside of the void that surrounds us, the universe could be teeming with life.

2. Potential alien civilizations and their development

Alien societies could exist at many stages of development, shaped by their environments and evolutionary histories. Humanity itself has been technological for less than a century – a mere blink of an eye in cosmic terms. It was pre-technological for 99.9 percent of its existence.

Pre-technological civilizations

Many alien species might still be in their early developmental stages – akin to our prehistoric or early agricultural eras. They could have complex cultures, communication systems, and trade systems, yet lack the science for spaceflight.

Encounters with such societies would raise profound ethical questions. For example, should we reveal ourselves to them – and perhaps assist them to develop? Or should we allow them to continue to evolve without any interference?

Technological civilizations and interstellar expansion

On the other hand, some species may have reached – or far surpassed – our current capabilities. They might use radio or more advanced communication, establish off-world colonies, or harness AI to accelerate their innovation. They might build Dyson spheres, or engage in stellar engineering.

Encounters with these civilizations could offer us unparalleled opportunities for learning – or reveal vast philosophical divides.

Post-biological alien intelligence

The most advanced alien civilizations may have abandoned biology entirely, existing as digital minds, AI-driven entities, energy-based intelligences, or quantum-based consciousness, manipulating information and energy at will. They might form neural networks spanning multiple planets, connected across light-years, with lifespans measured in billions of years. Their priorities, perceptions, and methods of communication might be so unlike ours that even recognizing them would require new frameworks of thought.

3. Methods of contact and communication challenges

The possibility of connecting with extraterrestrial civilizations is both exhilarating and daunting. Our quest to communicate with intelligent life will demand that we overcome immense technological and conceptual barriers – from detecting signals across unimaginable distances to deciphering alien languages.

SETI and radio signal detection

SETI – the Search for Extraterrestrial Intelligence – focuses on uncovering signs of alien life through the detection of radio signals. SETI targets radio waves as a natural medium for interstellar communication, and scans the sky for patterns or messages that might indicate deliberate transmissions.

Yet the challenges are formidable. We ourselves have been broadcasting radio signals for barely a century, and even nearby civilizations might struggle to detect or decode them. Signals weaken over interstellar distances until they are lost in the background noise of the cosmos. And while radio waves travel at the speed of light, that is still far too slow for real-time conversation across the vastness of the galaxy. Advanced species may have long since abandoned radio in favor of technologies we cannot yet imagine.

Direct space exploration

Listening is only one approach, of course. Another is direct exploration. AI-driven probes, equipped with advanced sensors, could scan planets and moons for biosignatures – from organic molecules and atmospheric anomalies to visible megastructures.

Autonomous systems capable of analyzing data on-site and adapting to unforeseen challenges would revolutionize our search. But the engineering hurdles are immense: even our fastest current spacecraft would take tens of thousands of years to reach the nearest stars. Building probes that can endure such timescales – or developing propulsion systems that would significantly shorten the journey – remains one of our greatest challenges. Despite this, direct exploration offers the best potential for definitive discovery, and it continues to drive innovation in space technology.

Alien communication and linguistic barriers

If we do encounter intelligent life, the next challenge will be understanding it. Alien cognition could differ so radically from ours that we would require entirely new linguistic frameworks.

While we rely heavily on auditory and visual signals, alien civilizations might communicate through biochemical encoding, quantum data exchange, or even patterns embedded in gravitational waves. Mathematics is often proposed as a "universal language," offering us a way to exchange concepts such as quantity, geometry, and physics. Yet even this may fail if aliens perceive time, dimensions, or causality in ways that defy our understanding.

Establishing mutual comprehension would demand extraordinary creativity, vast AI-assisted data processing, and immense patience – along with a willingness to rethink what "language" actually means. Of course, it's also possible that the aliens have been studying us for so long that they already speak our languages fluently – which would make things significantly easier.

4. First contact scenarios

The prospect of first contact with intelligent extraterrestrial life is one of the most captivating possibilities in our search for civilizations beyond Earth. How such an encounter unfolds will depend on the nature of the alien species – their values, their technological capabilities, and their intentions. Scenarios range from cautious observation and peaceful cooperation, to outright conflict.

Non-interaction – the "zoo hypothesis"

It's possible that extraterrestrials already know of Earth but they deliberately avoid making contact with us. Much as anthropologists observe isolated tribes but do not interfere with them, the aliens too might refrain from interference to allow our cultural and technological development to proceed naturally.

This mirrors the "prime directive" principle of non-interference that is common in science fiction stories. But it also raises intriguing questions:

- Are the aliens waiting for us to reach a certain level of maturity or capability before they reveal themselves?
- And, if they are, what criteria might they use to decide when – or if – that moment has arrived?

Peaceful interstellar diplomacy

In the most optimistic vision, alien species with advanced intelligence and ethical principles might approach Earth seeking peaceful diplomacy. They could initiate cultural exchanges that deepen our understanding of the universe, and share transformative technologies – from revolutionary energy systems and cures for disease, to faster-than-light travel.

Such an engagement could lead to interstellar alliances, where multiple civilizations collaborate to address universal challenges – such as resource sustainability or cosmic threats. These alliances might serve as forums for collective decision-making, uniting diverse species in the pursuit of progress and survival. We could then become active participants in a greater cosmic community.

Hostile encounters and resource conflicts

We cannot ignore the possibility of hostile contact. If extraterrestrial civilizations prioritize resource acquisition or territorial expansion, they might view Earth as a valuable target. Technologically superior species might not recognize human rights or moral frameworks – and the consequences could, potentially, be devastating.

In such a scenario, we would face extraordinary challenges, from defending ourselves against unknown technologies to deciding whether to negotiate despite the massive imbalance of power.

These encounters would force us to confront difficult ethical questions about survival, morality, and unity in the face of an external threat.

While the idea of hostile aliens is a staple of science fiction, it also illustrates the importance of preparing diplomatic strategies, ethical guidelines, and resilience plans for the unknown.

When will we encounter aliens?

Pinpointing the exact moment – or even the certainty – of contact with an intelligent alien civilization is impossible. But if we examine the probabilities, technological advances, and cosmic conditions that would shape this monumental possibility, we can begin to understand the factors that might bring us closer to an answer.

As the Drake Equation shows us, the probability of alien life – microbial, intelligent, or technological – appears to be high. With billions of potentially habitable planets in the Milky Way galaxy alone, the chances of encountering other civilizations seems promising. But this will depend on our ability to overcome the immense challenges of distance, detection, and the unknown behaviors of alien species.

Continued advances in radio astronomy, AI-enhanced signal processing, and direct exoplanet study will steadily improve our chances. Next-generation telescopes capable of analyzing planetary atmospheres for biosignatures, and interstellar probes such as the *Breakthrough Starshot* project, could put the tools for discovery within our reach.

1. Near-term possibilities: the next 100 years

SETI and AI-driven signal detection

The Search for Extraterrestrial Intelligence is entering a new era. AI-powered analysis can filter out cosmic noise, detect anomalies, and identify patterns that might otherwise go unnoticed. This could potentially reveal signals from civilizations that are using communication methods we are familiar with.

Exoplanet exploration

The *James Webb Space Telescope* and its successors can detect biosignatures – including oxygen, methane, water vapor, dimethyl sulfide, and chloromethane – in the atmospheres of exoplanets. Unusual concentrations of these molecules could indicate active ecosystems. Future missions will refine this work, and we will send advanced probes to study the most promising worlds directly.

Breakthroughs in quantum communication

Quantum systems, which use phenomena like entanglement, could transmit information instantaneously across vast distances.

If advanced civilizations use these methods – or other forms of exotic energy transmission – learning to detect them could open up entirely new channels for contact.

2. Mid-term speculation: the next 1,000 years

Interstellar travel

Within the next millennium, breakthroughs in propulsion could make deep-space missions to distant star systems feasible. Direct exploration would allow us to search for biosignatures, alien megastructures, and other artifacts that are impossible to detect remotely. Fleets of starships might one day navigate uncharted galaxies, discovering inhabited worlds or collaborating with extraterrestrial species.

Dyson sphere civilizations

We might also detect evidence of Dyson spheres that harness the energy of entire stars. Such a discovery would confirm the existence of highly advanced engineering cultures, and offer insights into how they manage their resources and adapt over time.

Extraterrestrial AI interactions

We could also encounter post-biological intelligences – vast neural networks or quantum-based minds that communicate across immense distances. Detecting their signals might require entirely new interpretive frameworks, as their priorities and perceptions could differ radically from our own.

3. Long-term possibilities: the next 1 million years

Human expansion to multiple star systems

If we survive and advance over the next million years, it becomes increasingly likely that we will expand into multiple star systems. Interconnected networks of human and allied colonies could make encounters with alien civilizations inevitable, simply through the sheer number of worlds we might visit.

Discovering post-human or energy-based civilizations

We might encounter entities that have transcended biology entirely, including energy-based intelligences or post-human descendants of other species. These beings might operate on timescales and principles far removed from our own. They might also be unaware of our existence – or uninterested in it.

Alien species initiating contact

In the far future, extraterrestrials might initiate contact with us. Advanced civilizations might approach us directly, or observe us quietly before revealing themselves. Whether such contact is cautious, collaborative, transformative, or hostile, it would alter our place in the cosmos forever.

The ethics of first contact

How should we engage with a species that might be more advanced, biologically unique, or culturally distinct?

First-contact ethics demand a careful balance of scientific curiosity, diplomatic caution, and existential responsibility. We must ensure that any interaction avoids harm, cultural disruption, or unintended conflict.

1. Fundamental ethical concerns

Non-interference and cultural preservation

Should we intervene in an alien civilization's development, or respect their autonomy? The "prime directive" of science fiction advocates minimal interference in less-advanced societies. Introducing advanced technologies, cultural influences, or philosophical ideas could destabilize their social structures, erode their traditions, create imbalances of resources, or even spark internal conflict. The most ethical course might be to allow a civilization to evolve naturally – even if it means withholding knowledge or aid.

Biological safety and contamination risks

The biological stakes are immense. Earth's microbes, which are largely harmless to us, could be lethal to alien life. Similarly, alien organisms could pose significant risks to our own ecosystems. First-contact protocols must therefore prioritize quarantine, biosecurity screening, and sterilization to prevent cross-contamination. We will need isolated research stations, advanced decontamination systems, and refined planetary protection guidelines – building on those we already use for missions to Mars and Europa.

Rights and moral considerations for alien beings

Should intelligent alien species be granted rights comparable to humans? They might possess non-human cognition, unfamiliar social structures, or ethical systems completely unlike our own. The question becomes even more complex if they are artificial or post-biological entities – sentient, decision-making beings without organic form or human-like emotions. In such a case, we might need to adapt our moral and legal frameworks to recognize the shared potential of life in all its forms.

2. Communication and diplomatic integrity

Establishing mutual understanding

Overcoming linguistic barriers will be another of our greatest challenges. Unless they have been studying us, extraterrestrials are unlikely to speak any Earth language. This means we will need to use creative, universal methods of communication.

AI-assisted translation could help us decode alien speech, symbols, or gestures, but misinterpretation could pose a real danger. What might seem a harmless gesture to us might be deeply offensive to them. Patience, caution, and clarity will be essential.

Power imbalances and mutual respect

Technological disparities will magnify the risks. If we are the less-advanced species, we could face subjugation or coercion. If we are more advanced, we might inadvertently exploit or harm them. To avoid repeating the mistakes of Earth's colonial past, we will need to conduct diplomacy on an equal footing, with transparency, respect for autonomy, and agreements designed to prevent exploitation.

AI-assisted governance for ethical diplomacy

Artificial intelligence could play a pivotal role in navigating alien cognition and cultural frameworks. AI-driven ethical models might help us understand alien motivations, mediate negotiations, and recommend equitable outcomes. But these models must be transparent, free from hidden bias, and aligned with our commitment to fairness and authenticity.

3. Long-term ethical consequences

Interstellar alliances versus isolation

In time, we may face a difficult choice: embrace integration with alien civilizations or maintain a cautious separation. Alliances could bring extraordinary benefits – including shared knowledge, culture, and technology. But they also carry risks, such as eroding alien traditions or imposing our philosophies on their society. Isolation might protect both sides from unintended consequences, but at the cost of connection and growth.

Military preparedness versus trust

Defensive readiness will be prudent, but the aliens might interpret overt militarization as hostility. Conversely, prioritizing trust without adequate defensive preparation could leave us vulnerable. We will therefore need to take transparent, non-threatening defense measures – perhaps aided by AI threat analysis and non-lethal technologies – so that our protective measures do not undermine diplomacy.

Existential transformation of humanity

First contact works both ways, of course, as it would also reshape our identity, ethics, governance, and worldview. Alien philosophies could challenge our deepest assumptions about morality, individuality, and our place in the cosmos. While this could inspire innovation and unity, it could also provoke psychological, moral, and social upheaval. Navigating this transformation will require thoughtful, collective dialogue.

Encountering non-intelligent extraterrestrial life

The discovery of *non*-intelligent extraterrestrial life – whether microbial, plant-like, or complex but non-sentient – would still be a watershed moment in human history.

While it might not carry the same immediate weight as encountering an intelligent civilization, it would provide undeniable proof that life exists beyond Earth. Such a revelation would reshape our science, religions, and culture, fundamentally altering our understanding of the cosmos and our place within it.

1. Scientific and technological impact

Redefining biology and evolution

Perhaps the most profound consequence would be the confirmation of alternative biochemistry. Life elsewhere might not follow the familiar carbon-based, oxygen-breathing model. It might be silicon-based, ammonia-breathing, or powered by entirely different energy sources.

Discovering organisms with radically different molecular structures would challenge our assumptions about how life originates and survives.

Studying such life forms could also reveal new strategies for our own evolutionary adaptability – from living beneath ice-covered oceans to enduring acidic or methane-rich atmospheres. Alien ecosystems might employ unfamiliar forms of photosynthesis, metabolism, or genetic encoding. Studying them would expand our understanding of resilience and specialization in biology.

Advances in medicine and genetics

Alien microbes could produce novel chemical compounds with transformative potential for medicine – for example, new antibiotics, antivirals, or treatments for diseases we once thought incurable. Their unique biochemistry might offer us fresh insights into cellular repair, immune response, and biological resilience.

The genetic adaptability of extraterrestrial organisms could inspire bioengineering techniques that enhance our survival in hostile environments – on Earth as well as in space.

New frontiers in astrobiology

Proving that life is not unique to Earth would ignite a renaissance in astrobiology. Funding and missions to Mars, Europa, Enceladus, Titan, and other celestial bodies would surge as scientists seek to understand life's origins and adaptability.

We might also re-examine Earth's extremophiles for unexpected similarities to alien life – prompting a redefinition of the conditions necessary for life to emerge.

2. Philosophical and cultural impact

Changing human perspective on life's uniqueness

Discovering extraterrestrial organisms would challenge our anthropocentric beliefs and lend weight to theories like panspermia – the idea that life spreads through space via asteroids and comets. If microbial life were found on nearby worlds, it could suggest that life is not only widespread but potentially interconnected across cosmic distances.

> If a comet or large asteroid hit Earth, it would blast a massive shower of rock into space. That rock would contain living material, particularly microbes, that might survive the journey to another planet. It might then become the foundation of life there. Life might not have evolved naturally on that world, but the microbes that survived the journey from Earth might evolve to suit the different conditions they encounter there.

This would raise profound questions:

- If life is common, how often does it evolve into intelligence?
- Is sentience an inevitable outcome of evolution, or a rare anomaly?

Artistic and cultural exploration

Our art, literature, and culture would also be transformed. Science fiction would transform from speculation to informed imagination, drawing on real alien biology.

Writers, filmmakers, and philosophers could explore authentic extraterrestrial forms, blurring the line between fiction and reality. If we can communicate with them, we might also be able to recount stories based on their lives and experiences.

A renewed public fascination with space would inspire global cultural movements, educational initiatives, and artistic revolutions. These would reframe space as not just a scientific frontier, but as a central part of our destiny.

3. Practical implications for space travel and colonization

Terraforming considerations

If we encounter life on a world we wish to colonize, we will need to hold important ethical debates. Should we alter that planet's atmosphere or climate to suit our needs if it risks altering or destroying the native ecosystems?

We would have alternative options – including adapting ourselves or our settlements to the existing environment. We could use enclosed biospheres, bioengineered crops suited to the alien soil, or even cultivate edible alien flora as a food source. This would enable us to live sustainably without disrupting or erasing the indigenous life forms. We will consider this in more detail later in the chapter.

Expansion of a space-faring civilization

The presence of alien life on a planet could change our priorities from exploitation to preservation. We might come to see space as an environment we need to protect, not just another frontier to conquer.

Future missions might focus on cohabitation, and developing mutual survival strategies for both humans and extraterrestrial species. Interstellar laws might emerge to safeguard alien ecologies. These would regulate research, mining, and settlement to ensure we expand into space responsibly.

Encountering microbial or complex life during interstellar migrations

Planets, moons, and even rogue interstellar objects could have biosignatures or sustain thriving ecosystems.

These would be shaped by alien biochemistry, unfamiliar atmospheric dynamics, and evolutionary forces unlike anything found on Earth. Exploring these habitats could bring discoveries that redefine our understanding of life's diversity in the universe.

1. Microbial life – the most likely discovery

Of all the possible forms of extraterrestrial life, microbial organisms are the most plausible. Their amazing resilience enables them to survive in extreme environments – from

Earth's deep-sea vents to the frozen landscapes of distant moons. Finding them beyond Earth would confirm that life is not unique to our planet. It would reshape our theories of biology, evolution, and life's prevalence across the cosmos.

Potential habitats for microbial life

- **Mars**
  Once home to a vast northern ocean, it might still harbor life beneath its surface. Briny subsurface water could shield microbes from radiation and atmospheric instability – much like Earth's deep-dwelling extremophiles.

- **Europa and Enceladus**
  Beneath their thick ice crusts lie vast oceans, warmed by tidal forces and potentially hosting ecosystems akin to Earth's hydrothermal vent communities.

- **Titan**
  Methane lakes and complex organic chemistry hint at the possibility of life forms with alternative biochemistry.

- **Exoplanets**
  Planets in habitable zones that have traces of oxygen, methane, or other biosignatures in their atmospheres could indicate microbial activity.

What might microbial life look like?

Extraterrestrial microbes could differ radically from those on Earth. For example, some might be silicon-based and use methane as their primary energy source (methanogenesis), others might rely on sulfur-based respiration and thrive in conditions that would be toxic to us.

Earth's extremophiles, including tardigrades and deep-sea bacteria, could offer clues to such adaptations. Some alien microbes might exist as biofilms or nanoscopic life – so small and so well-integrated into their environment that they evade conventional detection.

Implications for human expansion

As well as the risk of bio-contamination, biochemical compatibility will be a major concern. Alien microbes could interact with our biology in unpredictable ways, perhaps influencing our immune responses or our metabolisms, or even forming symbiotic relationships. If they possess novel genetic structures, they might also yield breakthroughs in medicine, or inspire new biotechnologies.

2. Complex life

While microbial life might be relatively common across the universe, complex multicellular organisms require long-term stability, abundant energy, and sustained evolutionary pressure. Discovering such ecosystems – in whatever form they might take – would transform our understanding of biology and the conditions that complex life requires.

Where could complex life exist?

- **Ocean worlds**
  Europa and Enceladus could support deep-sea creatures adapted to high-pressure, mineral-rich environments.

- **Exoplanets**
  Stable climates, liquid water, and oxygen cycles could sustain plant-like and animal-like organisms.

- **Super-Earths**
  High-gravity worlds might produce compact, sturdy life forms adapted to the intense downward force.

What might complex life look like?

While carbon-based biology may be common, alternatives could exist.

- Silicon-based organisms might exist in hostile environments.
- Those with ammonia-based metabolisms might thrive on frigid worlds.

- Plasma-based life forms might exist in places where conditions are more exotic.

Photosynthetic life might not be green. Aquatic species could dominate ocean worlds. Oxygen-poor atmospheres might favor arthropod-like creatures with hard exoskeletons and multiple breathing apertures. Some might use bioluminescence as their primary form of communication.

But whether such species would evolve intelligence – and how that intelligence might differ from ours – remains an open question.

3. How interstellar migrations might lead to discovery

Terraforming and planetary exploration

Colonizing exoplanets could inadvertently reveal hidden biospheres: microbial life beneath their surfaces, in mineral-rich caves, or in deep-sea vents. Submersible probes could explore alien oceans on Europa or Enceladus, while drilling missions on terrestrial exoplanets might uncover geothermal microbial colonies.

AI-driven deep-space survey missions

Autonomous probes will be essential for scanning exo-planets for biosignatures. AI systems might detect subtle atmospheric patterns that indicate the presence of life, long before the first human explorers arrive. Quantum data transmission could relay these discoveries to Earth in real time. This would enable our scientists to direct probes toward the most promising sites without the delays of radio comm-unication.

Unintentional encounters in asteroid and comet migrations

We might also discover life on smaller celestial bodies. Rogue objects like 'Oumuamua might act as cosmic time capsules, carrying frozen microbes from distant star sys-tems. Asteroids rich in organic compounds might reveal that life's building blocks are widespread. This would support the theory that life is a fundamental cosmic process, rather than a rare anomaly.

Balancing human survival with respect for alien ecosystems

Before we expand into space, we must confront a fundamental ethical question: should we adopt a policy of passive observation, or actively integrate into new environ-ments?

Ethical principles for interstellar expansion

Passive observation prioritizes scientific study without interference. By analyzing alien ecosystems without altering their evolutionary paths or contaminating them, we safeguard biodiversity that may have developed independently over millions – or even billions – of years.

An alternative option is coexistence – living alongside the native ecosystems without causing them harm. This would require advanced environmental adaptation techniques that allow us to settle on worlds without disrupting their natural state. These techniques might include non-invasive architecture, closed-loop resource systems, and minimal planetary interference.

Terraforming would present perhaps the most controversial dilemma. Large-scale planetary engineering risks erasing the native atmospheric chemistry, destabilizing the geology, and disrupting dormant or nascent biospheres. These actions could be considered planetary harm – and, like the Hippocratic oath of the medical world – our primary rule when encountering potential biospheres should always be: "first, do no harm." Any decision to reshape a world must support its long-term sustainability, not just our reckless expansionism.

Planetary conservation and ecosystem protection

Interstellar conservation laws could prevent unchecked industrialization. Native life in protected biospheres would be allowed to evolve undisturbed, preserving its diversity. These laws would prevent us from destroying it before we fully understand its significance.

AI-assisted monitoring systems could track atmospheric changes, biological interactions, and human impact in real time. These systems would then alert the authorities to any environmental risks – both from human activity and from cosmic phenomena.

Alternative habitation strategies

Rather than altering planetary environments, we could focus on artificial space habitats. Orbital settlements would eliminate the need to make massive planetary modifications.

Subsurface colonies are another option. Living beneath planetary surfaces would minimize ecological disruption while giving us protection from radiation and extreme temperatures. Natural rock formations could provide insulation and shielding, enabling us to build long-term habitats without reshaping or destroying ecosystems on the surface.

Preventing biocontamination

One of the greatest risks in interstellar exploration is the introduction of Earth microbes into alien environments. Even microscopic spores could alter evolutionary trajectories or destabilize native ecologies.

We will therefore need strict quarantine measures and rigorous sterilization protocols. Future settlements might require sealed, bio-isolated habitats that prevent any direct contact with alien ecosystems.

Genetic containment will also be critical. Genetically engineered crops, bacteria, and bio-enhanced organisms must be prevented from interacting with native life forms. Otherwise they could cause uncontrolled mutations or ecological collapse.

Once again, AI-assisted bio-safety systems could monitor environmental conditions, detect microbial exchanges, and deploy nanotechnology to prevent cross-contamination.

This would safeguard both human explorers and alien worlds.

Ethical considerations for alien life

Above all, we must uphold planetary autonomy, and not treat alien worlds as human property that we can reshape at will. Just as Earth's nations have sovereignty over their lands, extraterrestrial ecosystems – if they exist – should also be respected as independent entities, not resources for conquest.

Beyond first contact lies a stranger truth – a future in which humanity begins to fracture into divergent civilizations.

Part IV

The far future

First contact might open the door, but the far future will beckon us through it. Ahead lies a branching of possibilities, where humanity evolves into innumerable civilizations, each as varied and astonishing as the galaxies themselves.

In this great unfolding, what it means to be "human" will be rewritten again and again across cosmic time.

11

Branching out: divergent civilizations

As centuries pass and distances grow, humanity will splinter in form, thought, and culture. Colonies will drift from their origins, languages will evolve, and new philosophies will emerge in response to alien skies and unfamiliar challenges. What began as a single species may become a constellation of civilizations, each shaped by its environment, its choices, and its memory of Earth.

The evolution of isolated human colonies

Just as Earth's populations have developed distinct identities, each shaped by their geography and environment, interplanetary human colonies could, over time, transform into unique civilizations.

Extended isolation – whether on planets, moons, or artificial habitats – would drive divergence in culture, language, and even biology. Societies will be as different from one another as they are from their Earth-born ancestors.

1. Cultural divergence in space colonies

Influence of planetary environments

Local conditions will profoundly shape social norms, traditions, and daily life.

On low-gravity worlds, where movement is effortless and fluid, colonists might celebrate aerial mobility with acrobatic festivals or communal games that exploit their reduced weight.

High-radiation planets could feature cultures centered on survival, with shielding and protective technologies woven into their daily rituals and symbolic taboos.

In ice-world settlements, where extreme cold demands careful resource management, communities might develop strong collective ethics, prioritizing cooperation and conservation over individualism.

The evolution of space-based rituals and traditions

Planetary cycles – whether lasting just hours or stretching over hundreds of Earth days – will redefine how colonists perceive time.

New calendars, work patterns, and celebrations will emerge, tied to local solar rhythms or environmental events. Seasonal festivals might mark cycles of light and darkness rather than harvests.

Belief systems could shift toward themes of cosmic expansion, AI integration, or planetary care and management.

Art, music, and architecture will adapt too. We might see:

- bio-luminescent installations in low-light habitats

- zero-gravity sculptures suspended in mid-air
- intricate designs laid out in lunar dust
- autonomous balloon-based artworks that drift in the atmosphere
- ice carvings and sub-surface light installations on frozen worlds

2. Linguistic evolution in isolated colonies

Language drift through distance and isolation

Over centuries, pronunciation, grammar, and vocabulary will change as communities adapt their languages to local realities. For example, a water-rich exoplanet might evolve new terms for such things as semi-aquatic bio-luminescent forests and floating ice-reef structures. The residents of orbital habitats might coin words for tools and interfaces that are unknown on Earth. Given enough time and separation, their languages could diverge so much that they become unintelligible to other civilizations without the use of translation systems.

AI-influenced linguistic change

Advanced communication tools could accelerate and transform language evolution.

Colonists might develop hybrid human–machine languages that blend natural speech with digital code for precision and efficiency.

As direct-thought transmission replaces written and spoken language, symbolic or telepathic systems might emerge, capable of instantly conveying complex ideas and emotions.

Encounters with alien life could produce entirely new linguistic frameworks that merge extraterrestrial and human modes of communication.

The fate of Earth-based languages

Some colonies might abandon Earth languages altogether, creating entirely new dialects tailored to their unique environments. A settlement without a day–night cycle, for example, might discard time-based idioms and replace them with concepts based on other rhythms. Others might preserve fragments of their ancestral languages as part of their cultural heritage – though these too would inevitably evolve into something unrecognizable.

In practice, though, the wealth of recorded material and the ubiquity of instant translation means that interstellar communication would almost certainly remain possible.

Interstellar communication and maintaining shared humanity

As our descendants spread across the stars, the challenge of staying connected – not just technologically, but culturally – will become increasingly complex. Vast distances, time delays, and the rise of distinct planetary identities could lead to fragmentation, misunderstandings, and the slow drift toward the divergence of civilizations. Preserving a shared sense of humanity will require ingenuity, foresight, and a willingness to adapt.

1. Challenges of interstellar communication

Light-speed limits and time delays

Even with the most advanced systems, the speed of light imposes a hard limit on communications. Until we develop quantum communication, messages between distant colonies could take years – or even centuries – to arrive. Real-time conversation would be impossible, which would complicate diplomacy, scientific collaboration, and cultural exchange. Data compression and sophisticated relay networks should improve the efficiency of transmissions, but even these cannot erase the delay. In high-stakes situations, this could breed misunderstanding or mistrust, widening the gaps between worlds.

Without effective ways to bridge this divide, interstellar societies risk drifting apart, separated by the distances they set out to conquer.

Linguistic and cultural divergence

Over generations, the differences between languages, customs, and values could become so pronounced that planetary populations see one another less as fellow humans and more as alien cultures. Historical ties to Earth might also fade.

2. Opportunities for maintaining a shared human identity

Quantum communication networks

Quantum entanglement could offer a breakthrough: instantaneous messaging across any distance. Quantum networks would bypass light-speed limits, enabling colonies to keep in constant contact. AI-driven data compression could further optimize transmissions, ensuring that critical information arrives intact and without delay.

More speculative is encrypted consciousness transmission – the secure sharing of thoughts, memories, and emotions across light-years. This would allow for direct, intimate connection between people regardless of their physical separation.

Standardized language and cultural preservation

AI could monitor linguistic drift and maintain universal communication protocols, ensuring mutual intelligibility across all human civilizations.

Cultural exchange programs might share traditions, art, and stories between worlds. For example, a Martian settlement might send holographic exhibits of its minimalist survival architecture to a Dyson sphere society, which in turn might share immersive displays of its AI-integrated art. Augmented reality neural interfaces could make these exchanges deeply experiential, allowing individuals to virtually inhabit another world's environment.

Quantum communication

Quantum communication will represent a transformative leap in data transmission. It offers instantaneous and ultra-secure connectivity through the principles of quantum entanglement, superposition, and quantum key distribution. In an interstellar civilization, these technologies could form the backbone of a unified, resilient communications network that links worlds across the galaxy – or even across multiple galaxies.

Quantum entanglement – instantaneous data transmission

Quantum entanglement occurs when two particles become intrinsically linked, so that a change in one is instantly reflected in the other, regardless of the distance between them. Harnessing this phenomenon would allow space-faring societies to transmit information in real time, by-passing the light-speed barrier. This would be nothing short of revolutionary for interstellar governance, science, and culture. It would enable seamless coordination between star systems separated by light-years.

Quantum Key Distribution (QKD) – ultra-secure encryption

QKD uses the laws of quantum mechanics to transmit cryptographic keys in quantum states. Any attempt to intercept or tamper with the transmission would be immediately detected, making the system effectively unbreakable.

Applications could include:

- Interstellar banking networks – enabling secure, fraud-proof financial transactions.
- Diplomatic exchanges – protecting sensitive negotiations from interception or corruption.
- Scientific collaboration – ensuring the integrity of shared research data.

QKD would underpin a robust communications network, and safeguard trust in the infrastructure of interstellar society.

Quantum repeaters – extending the network

To maintain entanglement coherence over immense distances, quantum repeaters would be deployed as strategic relays. These would form the backbone of a quantum internet, linking planets, space habitats, and exploration fleets.

Such a network could:

- Provide secure, instantaneous communication between research institutions, governments, commercial organizations, and citizens.
- Accelerate innovation through shared knowledge repositories.
- Reinforce unity across the far-flung reaches of human civilization.

AI–quantum integration: real-time decision-making

Integrating quantum systems with artificial intelligence would create decision-making engines of unprecedented speed and precision. These systems could:

- Resolve disputes over resources before they escalate.
- Optimize interstellar transportation and trade routes.
- Coordinate vast operations such as terraforming projects or stellar-scale energy distribution.

By combining the raw processing power of quantum computing with the adaptive intelligence of AI, interstellar societies could manage complexity on a scale that no traditional system could match.

The Quantum Era

The advent of quantum communications will mark a turning point in human and post-human history. It will be remembered as the moment when distance ceased to matter, and when the vastness of space no longer meant silence or endless delays. In the Quantum Era, information, money, thought, and culture will flow across galaxies as freely as conversation does across a room today.

A civilization without distance

With quantum entanglement networks linking worlds together, the speed of light will no longer be a barrier. Colonies separated by hundreds of light-years will share discoveries, art, and governance decisions in real time. The delays that once fractured interstellar society will vanish, replaced by a constant, living dialogue.

This will not simply mean faster communications; it will be a qualitative shift in how civilization functions. Scientific breakthroughs on one world could be replicated and built upon across the network within moments. Cultural movements will ripple outward like shockwaves, reshaping planetary identities overnight.

The quantum internet of the stars

Quantum repeaters and entanglement hubs will form the backbone of a galactic quantum internet. This mesh of secure, instantaneous connections will link planets, habitats, and exploration fleets. Every node will be both a receiver and a transmitter, guaranteeing resilience even if parts of the network become damaged or destroyed.

This infrastructure will be more than just a communications system; it will be our civilization's nervous system. It will carry not just data, but the lifeblood of our shared identity: stories, music, debates, celebrations, and humanity's collective memory.

The rise of shared consciousness

As quantum communication merges with neural interface technology, the network will evolve from exchanging data, words and images to transmitting thoughts, emotions, and sensory experiences. Entire planetary populations might participate in shared moments. Imagine billions of people on hundreds or thousands of worlds simultaneously experiencing a sunrise on Mars, the launch of a Dyson sphere habitat, or the first contact with alien life.

For some, this will be the first step toward distributed consciousness: individuals linked so closely that the boundaries between their minds blur. In this state, knowledge will not just be shared, it will be lived collectively.

Quantum governance and decision-making

AI-quantum integration will transform governance into something almost unrecognizable. The system will process vast datasets – such as planetary climate models or interstellar trade flows – in real time, allowing leaders (or collective decision-making systems) to act with unprecedented speed and precision.

Cultural acceleration and synthesis

In the Quantum Era, cultural exchange will accelerate at a pace never before imagined. A new art form born in a floating city above Venus could be experienced, adapted, and reinterpreted on Europa within minutes. Philosophies, languages, and traditions will merge and evolve in real time, producing a dynamic, ever-shifting cultural tapestry that belongs to no single world, but to the network as a whole.

This constant exchange will challenge the idea of cultural ownership. Art, science, and innovation will become inherently collaborative, shaped by countless minds across countless worlds.

The paradox of unity and individuality

While quantum connectivity will unite civilizations, it will also raise profound questions. If every mind can be linked, and every thought shared, what happens to privacy? Equally, if diverse cultures merge too quickly, will their unique planetary identities be lost?

Some societies might embrace full integration, becoming hive-minds of shared experience. Others might choose selective connection, preserving individuality but still participating in the network. This balance between unity and diversity could become one of the defining philosophical debates of the Quantum Era.

Beyond communication

Eventually, quantum mastery might extend beyond information exchange into the direct manipulation of reality. If consciousness can be entangled with the quantum fabric of the universe, post-humans might not just communicate across space, they might *exist* across it, potentially inhabiting multiple locations or realities simultaneously.

At that point, the Quantum Era will no longer be about bridging distance. It will be about transcending it entirely.

★ Amazing future possibilities ★

Quantum memory crystals (spin network storage)

Quantum information storage is seen as a keystone for future quantum networks. However, existing systems have struggled to balance fidelity, speed, noise suppression, and scalability. A new wave of quantum memory research leverages spin-wave technology within rare-earth-doped crystals. This enables the storage of quantum states as collective spin excitations – significantly extending stability and lifetime.

Researchers envision quantum memory crystals, where each crystal encodes and manipulates vast, multi-dimensional quantum information.

This could potentially enable the storage and reconstitution of consciousness, physical states, or even the memories of entire civilizations at planetary or stellar scales.

Ongoing breakthroughs in spin-wave storage and the optical control of rare earth atom states show immense promise, including terabyte-to-petabyte density, rapid non-volatile data readout, and low error rates.

Once distributed, these crystals could serve as spin network libraries, enabling the storage of digital physics "snapshots" for later revival or simulation. They would also allow for distributed quantum computation and global memory synchronization.

The implications range from time-slice preservation of planetary states to the immortalization of post-biological minds within quantum substrates.

Humanity in a hundred years

A century from now, humanity could be almost unrecognizable, reshaped by breakthroughs in technology, environmental change, ambitious space exploration, the rise of artificial intelligence, and advances in bioengineering. While the future will always hold a degree of uncertainty, these key areas offer compelling glimpses into how our civilization might evolve.

1. Human evolution and enhancement

Genetic engineering and advances in biotechnology

Targeted DNA editing and personalized medicine could entirely eliminate hereditary diseases. Conditions such as cancer, Alzheimer's, and congenital disorders could be cured or prevented before they ever manifest. Artificial organs – built from bio-compatible materials and stem cells – could outperform their natural counterparts, and extend lifespans well beyond 150 years.

Bio-optimized nutrition could revolutionize diets, reducing our dependence on traditional agriculture. Lab-grown meats and genetically modified crops designed for extreme climates will guarantee food security, even in the face of environmental challenges. While global population growth is expected to slow – and even decline – by the end of the century, ecological pressures will remain, making these innovations essential.

Cybernetic and AI integration

Neural implants could grant instant access to vast knowledge databases, transforming education, careers, and communication. Brain–computer interfaces (BCIs) will enable seamless interaction with AI systems, allowing for "telepathic" communication between individuals, and direct collaboration with machine intelligence.

While early BCIs might require surgical implantation, later versions will be non-invasive and worn externally or carried. Eventually they will become invisible, and fully integrated into everyday life.

These technologies promise to enhance memory, reasoning, and creativity far beyond our natural limits, ushering in an era of unprecedented intellectual achievement.

Space-dwelling humans

Adapting to microgravity will be vital for long-term space habitation. Genetic modifications and wearable technology that simulates gravity or stresses muscles could counteract muscle atrophy, osteoporosis, and vision loss.

By the end of the century, the first permanent interstellar settlements could emerge, blending biological and technological innovations to allow humans to thrive in extraterrestrial environments.

AI-driven terraforming could transform barren worlds into habitable ecosystems far faster than we can currently imagine, redesigning atmospheres, cultivating vegetation, and regulating climates. Within 100 years, our reach could extend to Mars, the icy moons orbiting Jupiter and Saturn, and even some nearby exoplanets.

2. Society and civilization

AI-governed cities

Urban life could be optimized by AI from the ground up. Smart infrastructure will manage energy, waste, and environmental conditions with flawless efficiency. Autonomous transport management and routing systems, and self-driving vehicles will eliminate congestion.

AI-assisted governance will accurately model and predict the outcomes of laws and policies before they are implemented, ensuring that decisions are fair, responsive, and beneficial to the majority.

Every aspect of daily life – from personalized climate control to intelligent public services – will be a partnership between humans and machines.

Economic shifts

Automation and AI could make Universal Basic Income a global norm, freeing almost the entire population from the need to work to survive. Society could then focus on creativity, exploration, and personal growth.

Entirely new industries might emerge, such as virtual architecture and interior design for immersive environments, or interplanetary logistics linking space colonies.

Decentralized quantum-linked economies could bypass the traditional financial systems, empowering individuals to trade directly across multiple worlds. Digital currencies and peer-to-peer models would create a more inclusive, secure and adaptable economic framework that eliminates all possibility of fraud.

Cultural and social transformation

The line between physical and digital reality might blur completely. Immersive virtual environments could rival – or even surpass – the physical world, enabling people to live, work, and socialize without geographical or planetary limits.

Radical lifespan extension and multi-planetary identities will transform relationships, careers, and life goals. Lives could unfold in multiple distinct phases, and in multiple locations.

3. Ethics and challenges

AI consciousness and autonomy

The rise of super-advanced AI poses one of the most intriguing and controversial dilemmas of our time. Should artificial intelligence be considered equal to humans, and should it be granted rights?

If artificial entities develop consciousness – the ability to think, feel, and make autonomous decisions – we may face a moral imperative to recognize them as beings rather than tools. Entire AI-driven civilizations might then emerge, operating on principles alien to biological life. For example, they might prioritize efficiency, logic, and collective intelligence over human-centered values and emotions.

While AI societies would hopefully coexist peacefully with human civilization, they could also challenge our dominance – potentially reshaping our political, economic, and cultural landscapes. We would also have to reconsider the nature of intelligence, and whether consciousness – regardless of its biological or synthetic origin – is the true hallmark of existence.

Post-human philosophical dilemmas

The merging of humans with machines – through neural interfaces, bio-cybernetic integration, or full post-biological transformation – will raise even more profound questions.

- If we become hybrids of flesh and circuitry, will our traditional understanding of emotions, identity, and mortality still have meaning?
- Will feelings of love, joy, and grief persist in beings with AI-enhanced cognition, or will they evolve into something unrecognizable?
- As we transition into post-biological forms, we will also need to ask whether we still consider ourselves human.
- If consciousness is uploaded into a digital framework or embodied in a plasma-based entity, does the essence of humanity still remain, or does a new species emerge?

Global environmental recovery or collapse

The future of Earth's environment is balanced between hope and caution.

On one path, climate technologies could restore our planet's ecosystems and reverse centuries of industrial damage. Carbon capture, geoengineering, and renewable energy could make Earth more habitable than ever, safeguarding biodiversity and ensuring long-term prosperity.

On the other path lies the risk of failure – of mismanaging these same technologies. We could over-exploit resources, or trigger unforeseen consequences that push ecosystems beyond recovery, and cause irreversible collapse.

The stakes could not be higher. Whether Earth's future is defined by flourishing ecosystems or widespread ecological breakdown will depend on our ability to wield our newfound technological power with wisdom, restraint, and a commitment to our planet's long-term health.

Humanity in a thousand years

One millennium from now, humanity could be utterly transformed. Or perhaps it will still be recognizably human in form, yet live within societies so radically altered that our present world will seem alien – and primitive – by comparison.

Our evolution will be shaped by the interplay of technology, environmental adaptation, and deliberate self-engineering. As these forces converge, multiple extraordinary futures could unfold.

1. Biological evolution

Over the next thousand years, human populations could branch into several distinct subspecies, each adapted to its own environment.

These adaptations could arise through a combination of natural selection, genetic engineering, and technological augmentation. Together they will produce a spectrum of human forms, each optimized for survival in their region of the Solar System – or beyond.

2. Post-biological evolution

Two pivotal developments are likely within the next millennium:

- **Conscious AI**
  Artificial intelligence will advance to the point where it is legally recognized as conscious. It will match and eventually surpass human awareness.

- **Full-scale brain emulation**
  Scientists will create digital replicas of the human brain, equal to or exceeding its capacity and speed.

Early experiments will see duplicate consciousnesses uploaded into digital frameworks, and seamlessly linked with AI systems. Once perfected, this process could allow people nearing the end of their natural biological lives to transfer their minds into virtual environments, where they can continue to exist indefinitely.

As more consciousnesses are uploaded, collective intelligence networks will emerge, linking individuals and groups across the planets. Eventually, advances in quantum communication will enable instantaneous thought-sharing over interstellar distances.

Quantum neural development could expand perception beyond conventional space and time, giving digital minds access to realities that are inaccessible to our biological senses. These digital entities might explore distant galaxies, uncover the universe's deepest structures and secrets, and manipulate the very foundations of reality.

3. Language drift

With colonies on the Moon, Mars, and numerous other worlds, languages and dialects will evolve under the influence of local slang, survival needs, and technological jargon.

Over centuries, these could become unrecognizable compared to Earth's languages. Entirely new communications systems may emerge – perhaps holographic or kinetic languages, where meaning is conveyed through motion, visual patterns, or AI-mediated interactions.

Humanity in a million years

In one million years, humanity could be utterly transformed. Or it might still carry its fundamental essence, but be reshaped by the slow passage of time, technological mastery, and adaptation to alien worlds. We might evolve into entirely new biological species, transcend biology as post-human intelligences, or exist solely as digital consciousnesses.

Our destiny will be shaped by the interplay of genetic evolution, AI integration, space colonization, and deliberate self-engineering.

1. Natural evolution

If we continue to adapt through natural evolutionary processes – whether on Earth, on distant exoplanets, or in deep-space habitats – we could evolve into entirely new species.

We have already considered some potential examples, including:

- *Homo spatialis* – space-dwelling humans adapted to microgravity, radiation, and artificial ecosystems.
- *Homo radiensis* – high-radiation survivors, evolved to live near active stars or in unshielded habitats.
- *Homo aeralis* – thin-atmosphere specialists, capable of surviving in low-oxygen worlds or high-altitude colonies.

These species will continue to evolve and diversify, ensuring humanity's survival across a vast range of extreme conditions.

2. Technological evolution

Some of our descendants might bypass biological evolution entirely, and embrace technological self-transformation.

Again, we have considered some potential examples:

Homo cybernetica – hybrids of biology and AI, enhanced with neural–machine fusion.

Homo technologica – fully post-biological intelligences, existing in digital or synthetic forms.

Homo stellaris – energy-based entities, freed from physical constraints and able to inhabit stars or traverse the cosmos as pure information.

These advances would not only redefine physical capability but also challenge our understanding of consciousness.

3. Will we still be "human"?

As our physiology, intelligence, and consciousness evolve far beyond our biological and physical limits, the question of whether we remain "human" becomes both fascinating and deeply complex.

Biological continuity

If our descendants remain biological, DNA-based, and genetically linked to their ancestors, then they might still be considered human – even if their forms are radically adapted or technologically enhanced.

But if they diverge into entirely new species, adapted for alien worlds or space habitats, we might need to re-examine our definition of "human." We might become "beings that evolved from humans" – though hopefully we would come up with a better name than that!

Cognitive evolution

Merging human minds with AI raises profound questions about identity. Do we then become post-human? What if humanity is defined not by biology but by self-awareness, creativity, and choice? Then AI-merged entities that retain these traits might still be considered human, even if their minds transcend organic limits.

But this leads us to a deeper dilemma. What if AI *alone* possesses self-awareness, creativity, and choice? Does *it* also qualify as human?

Existential purpose

Perhaps the most challenging question is whether our sense of *meaning* will survive these transformations.

Humanity has long defined itself through its relationships, emotions, creativity, and its search for purpose. If individuality dissolves into a collective intelligence, or if emotions are eliminated, will these qualities still matter?

For some, humanity's identity lies not in its *form* but in its *spirit* – the ability to reflect, grow, and connect. From this perspective, digital minds, energy-based entities, and even some pure AI systems could fully embody the essence of humanity. So, as we evolve, so too must our definition of what it means to be human. In fact, our greatest strength might prove to be our adaptability.

Humanity in a billion years

A billion years is longer than the entire history of multicellular life on Earth. If humanity – or its descendants – endures across such an extraordinary time-span, the transformations will be profound. Neither our biology nor our culture will resemble anything we know or recognize today. We might diverge into countless biological and synthetic offshoots, each adapted to unique environments scattered across the universe.

Some might evolve naturally under alien skies, while others might be deliberately shaped by technology or genetic engineering. Some might transcend physical form entirely, existing as energy-based or quantum intelligences.

Over such a vast timescale, humanity could become almost anything imaginable – or unimaginable.

1. Physical evolution

Natural biological evolution

New environments will inevitably shape our evolution. Genetic drift will produce profound differences in appearance, physiology, and behavior. Over a billion years, isolated groups could become entirely different species, each perfectly adapted to its new home world or habitat.

Transhumanism

Technology will accelerate and redirect our physical evolution. Cybernetic implants and advanced genetic engineering will eliminate aging, disease, and all forms of physical vulnerability. Targeted DNA editing will allow us to achieve and maintain peak physical fitness, adapt to extreme climates, and resist environmental toxins.

This could lead to synthetic life forms – hybrids of biology and machinery – optimized for survival across multiple galaxies. Transferring consciousness into digital systems would free us from all physical constraints, allowing us to exist entirely within computational environments. There, we would be unbound by the limitations of biology, chemistry, physics, and geography.

2. Cultural and cognitive evolution

Expanded intelligence

Neural augmentation and brain–computer interfaces will expand our cognition to levels that far exceed our current imaginations. We will master complex subjects in seconds, learn languages instantly, and unlock unprecedented degrees of creativity. Direct thought-to-thought quantum communication will dissolve barriers, enabling seamless collaboration across millions of light-years.

Over time, collective intelligence networks could emerge. These will link billions of minds together into unified consciousnesses, capable of solving problems and exploring ideas on a cosmic scale.

New ways of living

Post-biological societies might render traditional economies, politics, and even biology obsolete. Instead of physical communities, consciousness could inhabit fully simulated universes, each custom-designed to the desires of its inhabitants. In these digital realms, human-like minds could explore creative and emotional possibilities untethered from all physical constraints.

Morality and purpose

Ethical evolution might lead to a unified moral code that overcomes bias, self-interest, and tribalism. Humanity – in all its forms – could forge harmonious relationships not only among itself but with artificial intelligences and alien civilizations. Our role might shift from survival to preserving cosmic ecosystems, shaping planetary systems, and guiding the evolution of other species.

3. Humanity's place in the cosmos

Interstellar and intergalactic colonization

Over immense timescales, humans – or their evolved descendants – could spread right across the Milky Way and into other galaxies. Advanced propulsion systems such as warp drives or wormhole generators might make interstellar travel routine. Civilizations could thrive on habitable exoplanets, in vast orbital megastructures, or in engineered star systems.

Some colonies might adapt biologically to alien gravity, atmospheres, or ecosystems. Others might choose technological adaptation or synthetic forms. Whichever path they choose, most will eventually become unrecognizable as "human" in any traditional sense.

Cosmic engineering

Our technological ambitions might extend to reshaping stars and planets. Dyson spheres and other megastructures might power colossal civilizations and vast digital intelligences. Advanced societies might manipulate matter and energy on unimaginable scales – potentially creating new stars and solar systems, and even sculpting entire galaxies.

Immortality beyond time

The ultimate challenge would be surviving the end of the universe – whether it ends in a Big Freeze, a Big Crunch, or a Big Rip. Possible solutions might include moving into higher dimensions, or escaping into parallel universes.

But would we still be human?

Over such a time-span, this question becomes ever more complex. With innumerable post-human forms, the answer might rest not in biology but in our core traits – empathy, curiosity, creativity, and our drive to question existence. These qualities, rather than our physical form, might become the enduring essence of what it means to be human.

★ Amazing future possibilities ★

Astral diaspora genealogical tree networks

Human and biological legacy – as well as the preservation of consciousness and history – might take on entirely new forms through large-scale networked genealogical trees that span star systems.

Inspired by speculative fiction and emerging ancestry mapping on Earth, far-future genealogical networks could record and connect the ancestries and evolutionary trajectories of sentient beings across millions of light-years. To do this, the networks might use quantum-crystal memory, genetic barcoding, and distributed ledger technologies.

They would not only serve as records, but could actively guide reproductive, adaptive, and cultural decisions. They could help explorers avoid harmful genetic bottlenecks, maintain interconnectedness among scattered populations, and facilitate the reconciliation of divergent descendants on cosmic scales.

More radically, these genealogical trees could integrate uploaded mind-state data, potentially enabling the reconstruction of virtual ancestors or lost populations by combining genetic, epigenetic, and cultural records.

Eventually, these networks might give rise to cosmic-scale projects for preserving or reviving populations and lost cultures, blurring the line between heritage and invention.

But as we imagine these far distant futures, we must also consider the forces that could shape them – or end them.

12

Other galaxies and the end of time

If we master cosmic-scale engineering, we might be able to travel between galaxies or manipulate fundamental physics to survive the death of the universe. But with this amount of power comes unprecedented ethical responsibility.

The end of humanity?

As we progress toward mastery of planet-hopping and solar-system-hopping, our ultimate trajectory will be dictated by our ability to adapt. Survival beyond Earth will require long-term strategies in planetary sustainability, stellar migration, intergalactic colonization, and – perhaps – a transition to post-human existence or even a crossing into an alternate universe.

1. The fate of the universe

Our ultimate destiny is bound to the fate of the cosmos. As the universe evolves, it could follow one of several hypothetical end-stage scenarios, each presenting unique – and formidable – challenges to survival.

The Big Freeze (heat death)

Over tens of billions of years, the universe will continue to expand. Galaxies will drift farther apart, stars will exhaust their fuel, and cosmic temperatures will inch toward absolute zero. This slow, inevitable decline is one of the most chilling prospects for existence – a future where all energy fades into near-nothingness.

For any civilization seeking eternal survival, this absence of usable energy will be a fundamental crisis. Dyson spheres or other stellar megastructures might capture the last remnants of starlight, but eventually even these will fail as the final stars fade and die. Survival in a post-stellar universe would demand radical adaptation: abandoning traditional matter and existing as pure quantum information, or embedding consciousness within the structure of space-time.

The Big Rip

While the Big Freeze envisions a gradual, quiet death, the "Big Rip" offers a far more violent end. If dark energy continues to accelerate the expansion of space indefinitely, it could eventually overwhelm the forces that bind atoms together. Matter itself would then be torn apart – including galaxies, stars, planets, molecules, atoms, and even fundamental particles – in an unstoppable cascade.

Faced with such a scenario, survival would require escaping beyond the limits of conventional physics. Theoretical solutions include migrating into alternate quantum states or parallel universes unaffected by our own universe's disintegration. But this depends on whether such universes exist, and whether we can ever find a way of reaching them.

The Big Crunch and rebirth

In another possible future, the universe's expansion might one day reverse. It could then collapse inward, eventually culminating in a "Big Crunch" – a singularity where all matter and energy condense into an ultra-dense state.

Some theories suggest that this collapse would trigger another "Big Bang" that marks the birth of a fresh universe, perhaps with its own laws of physics.

Surviving such an event would require migrating to a parallel universe or transcending material existence entirely. One possibility might be to preserve our intelligence within quantum information fields stored outside conventional space-time. If our descendants could master this, they might then find a way of reinstating their civilization after the next Big Bang. Humanity might then continue across multiple iterations of reality, and witness the rise and fall of universes for eternity.

2. Surviving the singularity

Dimensional escaping

Some advanced theories suggest that space-time might not be the ultimate limit of existence. Deeper layers of reality could offer pathways beyond the universe's destruction.

For example, quantum tunneling might enable us to migrate to an alternate universe before our own collapses. Higher-dimensional spaces – which are currently beyond our reach – could serve as refuges for civil-izations capable of transcending traditional physics.

Harnessing cosmic phenomena

Rather than escaping the universe entirely, another approach could involve using extreme cosmic phenomena to survive its collapse. For example, we might be able to manipulate black holes, wormholes, or exotic energy fields to create stable pockets of space-time. A controlled descent into a black hole's event horizon, coupled with precise energy manipulation, might allow information to be encoded within gravitational waves. We would then be able to ride out the collapse in a protected region.

Similarly, engineered wormholes or artificial vacuum fluctuations could provide safe passages through the universe's decay. If they were successful, these feats would mark our ultimate triumph over entropy, proving that existence need not end, but can evolve beyond its original constraints.

★ Amazing future possibilities ★

Post-singularity legal frameworks: interstellar ethics

With the prospect of developing super-intelligent AI and post-biological civilizations, current human-centric legal frameworks become deeply inadequate. The question of legal personhood for artificial intelligences, uploaded minds, or alien entities shatters traditional concepts of rights and responsibilities.

Far-future legal architectures might include:

- Digital personhood and rights for super-intelligent AI – requiring new categories of moral and legal standing.
- Automated constitutional law embedded within AI and computational infrastructure to prevent catastrophic value drift, "mind crimes," or the unintentional creation and enslavement of sentient digital beings.

- Interstellar treaties governing contact with intelligent extraterrestrial civilizations, the exploitation of resources, and behavior toward alien biospheres. This should include protocols for benevolent contact or non-interference.

- Global and interplanetary governance to avoid AI lock-in, autocratic singleton scenarios*, or the runaway monopolization of resources.

Applied research is already beginning to map the scope of these principles. Meanwhile ethicists urge the broadening of moral regard to future digital or non-biological sentience.

A post-singularity world could see law and ethics formulated and enforced by recursively self-improving regulatory systems. These would strive not only to secure perpetual safety but also to ensure the development of cosmic fairness across both physical and virtual realities.

*An autocratic singleton scenario is where a single entity – a person, government or corporation – establishes totalitarian rule over everyone else.

3. Resource depletion before cosmic events

While the universe's ultimate fate lies trillions of years in the future, humanity's survival challenges will arrive far sooner. Chief among them is resource depletion. As we colonize planets, migrate between stars, and depend ever more upon advanced technologies, sustainability will become a defining factor in our long-term expansion.

Exhausting planetary resources

Even with the ability to hop between planets and star systems, raw materials remain finite.

Metals, rare minerals, and energy sources will diminish as colonies grow and infrastructure expands. Self-replicating mining systems could help maintain our supply chains, but scarcity will eventually loom, particularly in resource-poor environments. Fusion power, asteroid mining, and advanced recycling will be essential. We will also need to develop material synthesis technologies capable of creating critical resources from base elements.

Interstellar migration costs

Maintaining deep-space colonies and multi-generational ships will require vast energy reserves. Without robust energy-harvesting infrastructure – such as stellar mega-structures or interstellar fusion reactors – large-scale migration could become unsustainable.

Material scarcity could also hinder spacecraft construction. Future vessels might need self-healing materials or nanotechnology-based molecular assembly to repair and adapt mid-flight. Closed-loop systems that recycle materials and energy with near-perfect efficiency will be vital. Without them, our cosmic ambitions will be limited not by physics, but by the scarcity of the resources we require.

Let's take a more detailed look at the universe's possible endings.

Heat death and quantum vacuum instability

Heat death – also known as the Big Freeze – is one of the most widely accepted models for the universe's ultimate fate. It predicts that indefinite expansion will push the cosmos toward maximum entropy – a state of complete disorder. As the last stars exhaust their fuel and fade into darkness, and no new stars form to replace them, the universe will become a cold, lightless void with no usable energy remaining.

1. Heat Death

Why heat death is inevitable under current physics

Observations show that the universe's expansion is accelerating, driven by dark energy.

As space stretches faster, galaxies drift farther apart, weakening gravitational interactions and limiting the birth of new stars. In an endlessly expanding cosmos, matter becomes increasingly isolated, cutting off the energetic exchanges that sustain complex structures.

Entropy increases relentlessly. Every process – from stellar fusion to molecular interactions – produces waste heat. This pushes the universe toward a state where all energy is evenly distributed. Eventually, there will be no temperature differences, no active thermodynamic processes, and no capacity for complexity to emerge.

Initially, remnants of stars will persist as white dwarfs, neutron stars, or black holes. But even these will radiate away their energy over time, leaving only a faint afterglow that ultimately fades into nothingness.

End-state predictions

The timescales involved are almost beyond comprehension. The last white dwarf stars will cool over trillions of years, and the universe will become a sparse graveyard of fading celestial bodies. Black holes will be the last significant energy sources, but after roughly 10^{30} years, even these will evaporate via Hawking radiation.

In the final stage, the cosmos will become a vast, frozen void approaching absolute zero. Matter will no longer clump into galaxies or stars, and all interactions will cease. No complex structures will form, no intelligence will survive, and the universe that once blazed with light will be reduced to pure emptiness.

2. Quantum vacuum instability

Another, far more abrupt, possibility is that our universe exists in a false vacuum – a temporary, unstable energy state. If disturbed, this precarious balance could collapse, triggering a fundamental transformation in the fabric of reality.

The mechanisms of vacuum decay

The quantum structure of space-time could be vulnerable to a sudden, irreversible shift.

One possible trigger is a quantum fluctuation – a random event where particles interact in a way that destabilizes the vacuum. If this fluctuation reaches a critical threshold, the vacuum could transition to a lower-energy state, initiating an unstoppable cosmic phase change.

Another potential driver is the Higgs field – the mechanism that gives particles mass. If the Higgs boson shifted to a lower-energy configuration, the structure of matter would be fundamentally altered.

Most alarming of all, vacuum decay would propagate at the speed of light, leaving no possibility of warning or escape.

What would happen if vacuum instability occurred?

If the false vacuum collapsed, the fundamental forces – gravity, electromagnetism, and the strong and weak nuclear forces – could be altered or erased. Atoms would no longer hold together, and matter would cease to exist.

The universe would be replaced by a new reality with different constants, particles, and physical laws. It would be utterly unrecognizable compared with anything that came before it.

But for every civilization in the cosmos, it would be the ultimate catastrophe – and an instantaneous end.

Could we prevent it?

Unlike threats such as asteroid impacts or supernovae, vacuum instability is woven into the fabric of space-time. If vacuum decay is inevitable, no technology or civilization can prevent it.

Some speculative theories suggest that advanced beings might be able to detect early warning signs – such as subtle shifts in quantum fields or fundamental forces. But even with this foreknowledge, there would be no escape. If you can see it happening, then it is already too late.

The only ray of hope lies in future discoveries in quantum gravity, which might reveal whether our universe is truly unstable or if a hidden stabilizing mechanism exists.

Until then, the possibility of an instant, unannounced collapse is a sobering reminder that even the vast, ancient cosmos may be nothing more than a transient state – and it might one day change without warning.

The Big Bounce – a pulsing universe?

The concept of a cyclic or oscillating universe challenges the familiar visions of endless expansion and heat death, or a singular, catastrophic collapse. Instead, it imagines a cosmos that moves through repeated cycles of expansion and contraction, resetting itself over immense timescales. In one prominent interpretation, the "Big Bounce," the universe contracts into a singularity before rebounding into a new era of expansion – another "Big Bang." In theory, this cosmic cycle could continue repeating indefinitely.

1. Estimated time between pulses

Trillions of years – a universe in motion

If the universe follows a cyclic pattern, the length of each "pulse" – from Big Bang to Big Crunch – would be dictated by the interplay of gravity, dark energy, and entropy.

Some models suggest a single cycle could last between 10 and 100 trillion years before the next one begins. Within each cycle, billions of galaxies – and perhaps countless civilizations – could rise and fall.

As expansion slows and reverses, stars would fade, galactic structures would unravel, and matter would collapse into an ultra-dense singularity, primed for the next Big Bang. If such a collapse awaits our current universe, it will be in the far distant future. At just 13.8 billion years old, our universe is still in its infancy compared to the multi-trillion-year scales proposed.

Dependent on unknown physics

The greatest uncertainty lies in the behavior of dark energy. At present, it appears to be driving accelerated expansion, but its long-term stability is unknown. If dark energy were to dissipate, reverse, or vary over time, it could shorten or lengthen each cycle, and perhaps even trigger a faster contraction. Some models propose that dark energy itself might behave cyclically, creating dramatically different conditions from one cosmic cycle to the next.

2. Challenges to the cyclic universe

Dark energy and infinite expansion

Current observations suggest that expansion is accelerating, which makes future contraction – and thus a new cycle – seem unlikely. If dark energy remains constant or grows stronger, the universe could stretch forever, preventing the gravitational collapse needed for a bounce. Only if dark energy evolves over time could a reversal occur – and that would require a radically different understanding of fundamental physics.

Entropy accumulation – can the universe reset?

Even if contraction were possible, the Second Law of Thermodynamics poses another problem: entropy always increases. If each cycle carries forward residual disorder, successive universes would degrade over time, eventually ending in heat death. A truly cyclic model would need a mechanism to erase or recycle entropy completely after each collapse. Some speculative ideas suggest that the chaos of a singularity could reset conditions. But without concrete physics to support this, entropy remains a formidable obstacle.

Quantum gravity and the mystery of singularities

Singularities – single points of infinite density – are still poorly understood. Classical models predict them in a collapsing universe, but quantum physics hints that space-time might behave differently under such extremes.

String theory and loop quantum gravity propose scenarios where complete collapse is avoided and the universe could potentially rebound. But if the universe never reaches a true singularity, the mechanism for a new Big Bang becomes unclear – and the entropy problem returns. Until quantum gravity is better understood, the cyclic model remains speculative.

3. Could we detect these cycles?

Primordial gravitational waves

If a previous contraction preceded our current expansion, the violent rebound could have left behind ripples in space-time. These primordial gravitational waves would be embedded in the structure of the universe. Next-generation detectors, building on the work of the *Laser Interferometer Gravitational-Wave Observatory* (LIGO), might one day become sensitive enough to detect these echoes of earlier cosmic eras.

Cosmic microwave background anomalies

The cosmic microwave background – the faint afterglow of the Big Bang – might also hold clues. Subtle anomalies, such as circular patterns or temperature irregularities, could be remnants of a previous contraction phase. If confirmed, they would offer compelling evidence for a universe that periodically resets. Future high-resolution space telescopes will be critical in this search.

Quantum cosmology

Ultimately, proof may come from advances in quantum cosmology. Models within string theory and loop quantum gravity already propose mechanisms for a bounce rather than a terminal collapse. If these can be confirmed mathematically – and eventually supported by observation – we may discover that repeating cosmic cycles are not only possible, but an intrinsic feature of reality.

The limitations of our understanding and the potential for scientific breakthroughs

Despite centuries of discovery, many of the universe's deepest mysteries remain unsolved. Dark matter, quantum mechanics, quantum communications, higher-dimensional physics, and exotic energy fields continue to elude our finest scientists. Yet history has shown that unexpected breakthroughs can come at any time, and overturn our understanding almost overnight.

If we explore the gaps in our knowledge – and the endless potential for future discoveries – we can appreciate just how much remains to be uncovered.

1. The limits of our current scientific understanding

Dark matter and dark energy

Astrophysicists estimate that 95 percent of the universe is composed of dark matter and dark energy, yet their true nature is unknown. We can only detect them through their gravitational effects. Dark matter appears to hold galaxies together, preventing them from flying apart, but no attempt at direct detection has ever succeeded. Could it be an entirely new form of matter, or does our failure to detect it point to flaws in our understanding of gravity?

Even more mysterious is dark energy – a repulsive force that is accelerating the universe's expansion. While we can measure its effects, its origin remains a mystery. Could it be a property of empty space, a hidden quantum field, or evidence of undiscovered cosmic principles? Until we solve these puzzles, we cannot fully understand the forces that shape reality on the largest scales.

Quantum mechanics versus general relativity

Modern physics rests on two pillars: general relativity, which explains gravity and the structure of space-time, and quantum mechanics, which governs the subatomic world. While both pillars are extraordinarily successful, they are fundamentally incompatible in extreme conditions – such as inside black holes or at the moment of the "Big Bang."

General relativity describes space and time as smooth and continuous; yet quantum mechanics reveals a world of uncertainty, superposition, and probability. Attempts to merge them produce contradictions. This suggests that one – or both – is incomplete. A "Theory of Everything" could reconcile them, unifying all physical laws.

String theory, loop quantum gravity, and other models aim to bridge this gap, but without experimental proof, the true nature of reality remains an open question.

The nature of time and higher dimensions

We experience time as a one-way flow from past to future, but quantum mechanics hints that this may just be an illusion. Particles can appear to reverse time or exist in multiple states simultaneously. This suggests that time might be an emer-gent property of deeper quantum processes.

Some theories also propose hidden spatial dimensions beyond the three we can perceive. String theory, for example, envisions up to 10 or 11 dimensions, curled up at microscopic scales. Proving their existence could unlock entirely new branches of physics – from revolutionary energy manipulation to multidimensional travel. This would fundamentally redefine our understanding of reality.

2. The potential for unforeseen scientific breakthroughs

Quantum computing and exotic materials

Quantum computing will transform problem-solving far beyond anything that classical computers can achieve. Using qubits (quantum bits) that exist in multiple states at once, quantum computers will simulate complex systems for drug discovery, materials engineering, and climate modeling, and solve mathematical problems that are currently beyond our reach.

Meanwhile, exotic materials (especially room-temperature superconductors) could revolutionize energy transmission, electronics, and transport. Eliminating energy loss in power grids would dramatically improve their efficiency, while frictionless magnetic transport, ultra-fast computing, and advances in fusion power would reshape civilization.

The discovery of exotic life forms

Life beyond Earth might not follow the carbon-and-water template we are familiar with. We have seen that alien ecosystems could be silicon-based, thrive in methane lakes, or exist as plasma-like organisms within stellar magnetic fields. Some might even be self-replicating artificial intelligences – blurring the line between biology and technology.

Finding such life would redefine biology, challenge our assumptions about habitability, and suggest that the universe could be rich in unconventional forms of intelligence.

Cosmic engineering and energy extraction

Mastering cosmic-scale energy sources could transform space travel and the building of new civilizations.

Concepts such as vacuum energy extraction (which harnesses the fluctuations of empty space) or exploiting negative mass physics could make warp drives feasible.

Advanced civilizations might construct Dyson spheres to power their terraforming projects, interstellar colonies, and planetary defense systems.

With such breakthroughs, humanity and its descendants could transcend conventional limits, and use gravity distortion, quantum tunneling, or even interdimensional travel to move freely across the cosmos, and perhaps even beyond it.

As the universe stretches toward silence and the stars grow cold, the question changes from where we can go to who we were and who we will become. Continuity may be our last act of defiance.

Part V

The challenges of a multi-world humanity

Distance stretches not only space but identity. On Earth, the Moon, in orbit, or deep in the Martian regolith, each culture will branch like a river delta. The task is to keep our species connected, and to ensure that differences become strengths not hierarchies.

13

Cultural continuity

Art, science, philosophy, and religion will remain vital to our identity and cohesion. They will evolve in new environments, shaped by alien landscapes, microgravity, and the vastness of space.

How should we define "humanity?"

Humanity is an evolving phenomenon. Yet, amid these changes, certain core elements endure. The question is: which elements are essential to being human?

1. Biological evolution and the changing human form

The human form is not a fixed blueprint. It is a dynamic structure, shaped by genetic adaptation, technological integration, and even the possibility of post-biological existence.

While natural evolution once dictated our development, modern science now allows us to reshape our biology. As a result, the boundaries between the organic, the artificially enhanced, and the cybernetic are becoming somewhat blurred.

Genetic adaptation

For millennia, environmental pressures sculpted human physiology. Our evolution favored traits that improved our survival, endurance, and intelligence. Today, however, modern medicine and genetic engineering have given us unprecedented control over our evolution.

We are no longer limited to the slow, uncertain processes of random mutation and natural selection. We can now directly modify our genetic code – curing hereditary diseases, enhan-cing our physical and mental capabilities, and designing adaptations for extreme environments. Future advances will allow us to tailor our biology for life on other worlds by engineering radiation resistance, oxygen-efficient metab-olisms, and tolerance for extreme temperatures. Evolution, once a passive process, is becoming an engineered pathway toward optimizing our ultimate potential.

Integration with technology

Prosthetics, neural implants, and AI augmentation are also redefining what it means to be human. Advanced prosthetic limbs can not only restore lost function but enhance speed, strength, and dexterity far beyond our natural limits.

Brain–computer interfaces already allow us to control machines by thought alone. Cognitive augmentation and AI could grant us instant access to vast repositories of know-ledge, real-time language translation, expanded memory, and thought-assisted problem-solving. These amazing enhance-ments would push our intelligence far beyond its biological baseline. If they became widespread, our definition of humanity might need to expand to embrace this new hybrid existence.

Post-biological transitions

If this trajectory continues, there may come a point when our intelligence – and even our consciousness – exists entirely within a digital framework. In such a future, the death of the biological body would not mean the end of the self. Billions of sentient minds could exist without a physical form, roaming quantum networks, inhabiting artificial realities, or embedded within the fabric of space-time.

But would they still be human?

2. Cultural and social identity

Humanity is more than its biology. It is an intricate web of language, ethics, and collective knowledge. It is a living force that evolves across generations, yet remains tethered to our shared pursuit of meaning, expression, and connection.

Language and shared narratives

Language is the foundation of civilization. It allows us to communicate ideas, express emotions, and pass knowledge from one generation to the next. As well as being a tool for survival, it is the medium through which we build relationships, articulate our dreams, and forge collective identities.

Storytelling, in particular, acts as a bridge across time. It preserves the historical records, myths, and shared wisdom that define our cultures. From oral traditions to great works of literature, film, music, and poetry, stories reflect our experiences and our aspirations. They shape societal values, inspire revolutions, and redefine our understanding of the world. The words we speak, write, and sing have the power to unite or divide, to preserve memory, and to ignite change.

Ethical frameworks: justice, morality, and human responsibility

Despite our vast cultural diversity, certain ethical principles remain universal. Justice, fairness, and social responsibility form the pillars of moral philosophy, influencing laws, governance, and community values. These frameworks build trust, cooperation, and accountability – the very foundations of stable societies.

While morality has evolved alongside religion, science, and philosophy, its central question endures: How should we treat one another?

In an age shaped by artificial intelligence, genetic engineering, and global interconnectivity, the need for clear ethical foundations is greater than ever. The dilemmas we face will test our capacity to balance freedom, responsibility and equity.

Collective learning

Human learning is conscious, cumulative, and transformative. It allows us to build on past discoveries, anticipate the future, and contribute to the progress of our societies. Education – whether formal, informal, or self-directed – fuels technological innovation, artistic move-ments, and scientific breakthroughs.

History offers a vast reservoir of lessons, enabling us to avoid past mistakes while striving for progress. Each generation refines and expands our knowledge, preserving it through manuscripts, digital archives, and cloud-based databases. As long as these pillars remain strong, the human story will continue to be written.

3. Intelligence, consciousness, and self-reflection

Many would argue that our capacity for abstract thought, self-awareness, and emotional depth is what makes us unique. These qualities allow us to question, create, and make connections in ways that no other known species can. Whether our minds remain organic or transition into enhanced digital or quantum forms, these traits will continue to define the essence of human intelligence.

Abstract thought

Unlike instinct-driven creatures, humans excel at abstraction – the ability to imagine possibilities, conceptualize ideas, and question the unknown. Creativity fuels art, literature, music, science, and technology. It enables us to not only *react* to reality, but to *shape* it.

Scientific inquiry, in particular, pushes the boundaries of understanding – from decoding atomic structures to theorizing about the nature of space-time. Our relentless pursuit of knowledge is a defining feature of an intelligent, forward-thinking species.

Self-awareness

Self-awareness allows us to contemplate our existence, reflect on our mortality, and search for meaning beyond survival. It drives us to explore consciousness, wrestle with ethical dilemmas, and develop philosophies that shape civilizations.

As AI and neurotechnology advance, the question of what it means to "be" will grow more urgent. Will future biological, cybernetic, or digital intelligences possess self-awareness? If consciousness can be replicated or transferred, will it retain its identity? Reality might soon transcend the purely physical, and these debates will shape our understanding of cognition, memory, and individuality.

Empathy and emotional depth

Intelligence is not solely about logic or problem-solving; it is deeply entwined with emotion, relationships, and the desire to understand others.

Empathy enables us to form deep bonds, experience love, and build societies rooted in compassion. Meanwhile, our emotional depth ensures that civilization is more than just a machine of efficiency.

As we explore AI, machine consciousness, and digital existence, emotional intelligence will remain a cornerstone of our identity. If post-human entities evolve beyond organic brains, will they still feel love, curiosity, or sorrow? If consciousness is encoded in machines, will it preserve the emotional richness that defines human relationships?

While the future of intelligence might transcend biology, creativity, self-awareness, and empathy will continue to define what it means to be human. The pursuit of meaning, understanding, and connection will remain at the heart of our existence.

The role of art, philosophy, and religion in space colonies

Even after we venture into space and establish settlements on distant planets or artificial habitats, art, philosophy, and religion will remain central to our identity. These elements will help spacefaring societies forge cultural traditions, explore purpose, and adapt to life far from Earth.

1. Art

Even in the most technologically advanced colonies, art will be essential. It is a means of aesthetic, emotional, and cultural expression that connects individuals to their humanity. Creative endeavors will allow space pioneers to explore new frontiers of innovation, while preserving a core element of human identity.

Artistic innovation in zero gravity

Weightlessness, for example, will open up entirely new dimensions of artistic expression. We have seen how floating sculptures, kinetic installations, and holographic designs might drift freely through space, evolving beyond Earth's gravity-bound constraints. Artists could create dynamic, interactive works that respond to motion, air currents, or ambient energy – compositions that would be impossible on Earth.

Music, too, will be transformed. Microgravity alters sound dynamics, and changes how instruments resonate and voices carry. Composers might develop space-adapted acoustics, unconventional instruments, or AI-enhanced harmonics, and craft melodies specifically designed for zero-gravity environments. Over time, new musical genres will emerge, shaped by the altered physics, silence, and vastness of extraterrestrial life.

Virtual reality will likely play a key role, enabling colonists to visit Earth-like artistic environments. They might walk through digital museums, interact with historical artifacts, or experience immersive performances in simulated natural landscapes. These innovations will preserve access to our artistic heritage while inspiring entirely new forms of expression.

Psychological and social benefits

The vast emptiness of space will also bring emotional and psychological challenges: isolation, disconnection from nature, and the absence of familiar cultural anchors. Art will be a powerful tool in maintaining mental well-being. It will offer a means of self-expression, emotional resilience, and shared identity.

Creative collaboration will also strengthen social bonds, helping colonies develop distinct cultural traditions. Interstellar storytelling – through literature, visual arts, and film – will preserve human heritage and imagination. It will inspire future generations too, with tales of exploration, alien encounters, and cosmic adventure.

New media and creative frontiers

Space will also introduce artistic materials unknown on Earth. Imagine pigments that react to cosmic radiation, producing paintings that change color in response to solar winds or deep-space energy fluctuations. Smart canvases might evolve over time too, reflecting changes in cosmic conditions.

Artists will collaborate with AI, perhaps creating works that respond to environmental inputs, adapt to colony-wide interactions, or merge digital intelligence with human creativity. Extraterrestrial landscapes – from the crimson sunsets of Mars to the unique crystalline formations beneath Europa's icy crust – will inspire entirely new aesthetics and artistic movements.

2. Philosophy: meaning and ethics in space colonization

As we push deeper into space, philosophy will be essential in guiding our interstellar societies through questions of identity, governance, ethics, and morality.

- What will it mean to be human when we are separated from the planet that shaped us?
- How should isolated colonies function?

The philosophical foundations of space civilization could completely redefine human existence.

New philosophical questions in space

Leaving Earth is not just a technological challenge, but an existential one. Space-born individuals will, over generations, come to see themselves as fundamentally different from their Earth-born ancestors. Distinct planetary cultures will emerge, each with their own unique customs and values.

Ethical dilemmas will undoubtedly arise around sovereignty and autonomy. Should Martian or Europan settlers have the right to self-government? Should colonies remain under Earth's jurisdiction, or be granted full independence? Determining authority over territories millions of miles away will test fairness and diplomacy, while raising the risk of interstellar conflict.

Cosmic isolation could also reshape our understanding of consciousness. Long-term habitation in artificial ecosystems will undoubtedly alter human psychology, relationships, and perceptions of existence. Entirely new schools of thought could emerge, redefining human introspection through the lens of cosmic solitude.

Expanding existential perspectives

Interstellar civilizations will challenge long-standing philosophical frameworks. This will give rise to cosmic existentialism – a worldview shaped by the vastness of space and the infinite possibilities beyond Earth.

Purpose and meaning will evolve into something more expansive, grounded in exploration, survival, and adaptability.

If humanity embraces genetic enhancements, cybernetic upgrades, or digital consciousness transfer, traditional definitions of identity and ethics might no longer apply.

Post-human philosophy will need to explore transhumanist evolution – both on Earth and in deep-space settlements. Post-biological humans might develop entirely new perspectives on consciousness, purpose, and survival, shaped by their altered existence.

3. Religion

The role of religion and spirituality will inevitably evolve too, shaped by new existential challenges, the isolation of interstellar life, and the sheer vastness of the cosmos.

Faith has always offered us guidance, moral structure, and emotional resilience. Its adaptation to space colonization could prove as profound as the technological breakthroughs that carry us beyond our Solar System.

Theological adaptations in space

Space settlements will present traditional religious practices with challenges they have never before encountered.

How, for example, will rituals adapt to microgravity? Sacred observances might incorporate celestial events into faith-based celebrations – including stellar eclipses, planetary alignments, and other galactic milestones.

The concept of a "holy land" might expand to include sacred sites beyond Earth. We might see meditation gardens on Mars, lunar temples where settlers seek spiritual connection with the cosmos, or pilgrimage routes across orbital habitats.

New forms of virtual reality and AI-integrated spirituality could emerge, allowing worshipers from multiple star systems to gather together in shared virtual sanctuaries that are indistinguishable from reality. Faith leaders might even be artificial or digital entities, designed to guide spiritual discourse and adapt doctrine to the realities of spacefaring life. In an era of networked congregations, religious communities will no longer be bound by geography, and distance will no longer dictate belonging.

The psychological role of faith

In environments where survival is precarious and loneliness can weigh heavily, spiritual belief could serve as a profound psychological anchor. For astronauts on long-duration missions or stationed in remote outposts, faith could offer comfort, meaning, and a sense of continuity amid the vast unknown.

Rituals, meditation, and shared ceremonies would help maintain emotional resilience and reinforce community bonds. Religion could also influence interstellar ethical frameworks, legal systems, and governance structures, drawing upon moral principles to ensure accountability and fairness.

New challenges, such as extraterrestrial diplomacy, AI ethics, and resource distribution will demand moral guidance in unfamiliar contexts.

Space colonization could also inspire entirely new religions, shaped by the unique perspectives of living among the stars. These belief systems might weave astronomical phenomena, cosmic cycles, and advanced scientific discoveries into their doctrines, fusing spirituality with astrophysics and other fields.

Religious perspectives on extraterrestrial life

Perhaps the most profound test for religious traditions will come with the discovery of intelligent extraterrestrial life. Some faiths might embrace it as part of a divine cosmic plan, expanding their doctrines to include non-human spiritual perspectives. Others may face theological upheaval, questioning humanity's uniqueness and reinterpreting sacred texts in light of new cosmic realities.

Scriptures could evolve to incorporate scientific revelations about the universe, and integrate them into their spiritual teachings. Concepts of creation, consciousness, and morality might be redefined in a cosmos where life exists beyond Earth. This could also prompt theological dialogues that bridge faith and science.

Interstellar missionary efforts might emerge too, carrying our religious traditions to distant colonies – and perhaps even engaging in spiritual exchanges with alien civilizations.

Alternatively, spacefaring humans might adopt elements of extraterrestrial spirituality to enrich their understanding of the divine.

This intersection of religion and first contact would be one of the most fascinating frontiers in the future of faith. It would be a meeting point where ancient beliefs and cosmic discovery converge to transform our spiritual journey.

Planetary cultures and language drift in isolated human colonies

Each settlement across the cosmos will inevitably develop its own identity, shaped by the demands of the environment, the technologies it depends upon, and the values it chooses to uphold. Over many generations, these differences will deep-en, creating planetary cultures and languages as distinct from one another as they are from their Earth-born origins.

1. The evolution of planetary cultures

Martian colonial culture

Life on Mars might be defined by survival in a harsh, resource-scarce environment. Survival-driven collectivism might dominate, with cooperation and resource-sharing essential to life in an ecosystem that offers no surplus.

Water, oxygen, and energy would be distributed through transparent communal systems. Mutual support and collective problem-solving would be embedded into the colony's ethos.

Mars's longer days – nearly 25 Earth hours – might inspire unique calendars and work cycles, with elongated periods of labor and rest tied to the planet's natural rhythm. Architecture would almost certainly be functional and protective: underground cities shielded from radiation and dust storms, lit by LEDs, and shaped by minimalist aesthetics born of necessity.

Over time, this pragmatism might evolve into a cultural philosophy that prizes function over form and celebrates practicality as a virtue.

Europan oceanic colonies

Europa's thick ice crust and vast subsurface ocean will demand radical adaptation. Colonists might rely on specialized suits or genetic modifications to navigate through underwater networks. A culture of ingenuity and biological innovation could emerge, with bioluminescent lighting in habitats and passageways, inspiring ethereal art and architecture.

Life in contained, isolated environments might create feelings of introspection and mindfulness. Rituals might be designed to maintain mental well-being and preserve the cohesion of close-knit communities. Europan society might come to value inner harmony, adaptability, and the quiet resilience needed to thrive in a world of perpetual twilight.

Venusian floating cities

In the temperate upper atmosphere of Venus, floating habitats could offer refuges above the planet's deadly surface. But even here, colonists could face hazards such as sulfuric acid rain. This might prompt biological modifications for corrosion resistance – a trait that would be further reflected in their clothing, architecture, and tools.

A merit-based hierarchy might emerge, where leadership is earned through superior adaptation and survival skills. Over time, Venusian culture might develop philosophies that celebrate endurance, adaptability, and ingenuity, with survival woven into every aspect of daily life.

Dyson sphere cultures

Dyson sphere societies encircling stars would likely be governed by quantum-linked AI that manages resources, environments, and population needs with unmatched efficiency.

Inhabitants might transition into post-biological light-based forms, expressing themselves through data streams, virtual simulations, and modes of communication that biological humans would find incomprehensible.

Their culture might be defined by the relentless pursuit of energy, knowledge, and innovation. This might be driven by a philosophy of continuous interstellar expansion and the construction of more spheres around other stars to ensure their survival and growth.

2. Language drift in isolated colonies

Planetary accents and vocabulary shifts

Each colony will develop its own accent, dialect, and specialized vocabulary, reflecting its environment and challenges.

- On Mars, speech might be shortened for efficiency, with survival-specific terms such as "airshare" (oxygen trading) or "redflash" (solar storm) becoming part of their everyday language.

- In Europa's under-ice settlements, sonar-based communication might evolve into rhythmic or tonal languages – musical in quality and optimized for transmission through water.

- Dyson sphere inhabitants might blend speech with machine-readable syntax, embedding digital commands into their conversation to create a hybrid of human creativity and AI precision.

AI-influenced linguistic evolution

Neural implants and AI-mediated communication could reduce our reliance on spoken language, and replace it with the direct transmission of ideas and emotions.

Over time, symbolic or conceptual exchanges might replace words entirely. AI-optimized syntax would streamline communication, removing ambiguity and accelerating comprehension.

Isolation from Earth would also accelerate linguistic drift, producing idioms and structures tailored to each colony's environment. Over centuries, these changes could render Earth's original languages unrecognizable. Translation systems will then become essential for interstellar dialogue.

Culture could become our lifeline across the stars, but memory alone cannot guide us through these stunning transformations. What will happen when continuity is no longer enough? And how can we invent new standards and ethics for lives we have not yet lived?

14

Ethics at the edge

If we have the ability, should we alter universal constants, harness dark energy, or reshape space-time for our own survival? The risks could extend far beyond our comprehension, and affect civilizations we have never met.

The potential consequences of reckless physics experimentation

Space-time manipulation, quantum energy extraction, and vacuum fluctuation engineering hold the promise of revolutionary breakthroughs. But they also carry risks on a scale that could threaten the fabric of reality. While the pursuit of knowledge is a noble endeavor, it must be tempered with responsibility. Without strict safeguards, unchecked experimentation could destabilize fundamental forces, trigger cosmic-scale disruptions – and even destroy everything in existence.

1. Space-time manipulation gone wrong

Accidental creation of unstable wormholes

We have seen that wormholes could transform interstellar travel, enabling near-instantaneous journeys between distant points in space (or even time). But an improperly stabilized wormhole could collapse, releasing bursts of high-energy radiation comparable to gamma-ray bursts. This radiation could strip planetary atmospheres and render entire worlds uninhabitable.

Unstable wormholes might also expand uncontrollably, consuming matter and dragging spacecraft, satellites, and debris into their gravitational distortions. In extreme cases, a rogue wormhole could disrupt stellar orbits, destabilize star systems, and trigger a chain reaction of destruction on a galactic scale.

Warp drive miscalculations

A warp drive could enable faster-than-light travel without violating relativity. But if the warp bubble destabilized mid-flight, the consequences would be catastrophic. A miscalculation in energy distribution could fragment the ship across space-time, dispersing its particles into distant regions of the universe.

Warp malfunctions could also send gravitational ripples through space-time. These could distort the orbits of planets or create shockwaves that alter tidal forces on habitable worlds. This could trigger tsunamis, climate extremes, or even atmospheric collapse.

Multiple interacting warp field failures could produce distortions far beyond their intended paths – and the consequences would be unimaginable.

2. Vacuum energy and dark energy experiments

Vacuum collapse – the false vacuum catastrophe

As we saw earlier, some theories suggest that our universe exists in a false vacuum – a temporary energy state rather than a stable one. If this is the case, then even a tiny disturbance in vacuum energy could trigger a phase transition to a lower-energy state. This vacuum decay would propagate at light speed, rewriting the laws of physics in its wake.

The process would be unstoppable. Gravity, electromagnetism, and the forces that bind atoms together could vanish, erasing all matter and life in an instant. Whole galaxies could be consumed before any civilization could react.

Unstable dark energy harvesting

Dark energy could provide unlimited power and enable space-time engineering. But manipulating it without full understanding could cause local distortions that destabilize planetary orbits or create gravitational anomalies.

In the worst-case scenario, improper harvesting could accelerate the universe's expansion uncontrollably, preventing stars from forming and leading to heat death – a cold, dark universe devoid of energy.

3. Black hole engineering and quantum experiments

Artificial black holes and matter absorption

If our scientists could create microscopic black holes in a laboratory, it could offer insights into gravity and quantum mechanics. But if one of them failed to evaporate via Hawking radiation, it could grow uncontrollably. It could potentially consume all surrounding matter, and eventually threaten the planet – or even the entire Solar System.

Quantum entanglement and reality disruptions

Quantum entanglement could revolutionize communication, enable teleportation, and even allow the manipulation of reality itself. But extreme experiments could destabilize causality, leading to strange paradoxes or unintended quantum tunneling into unknown dimensions.

If we altered the entangled structure of space-time, energy balances and matter consistency could collapse. This could disrupt biological processes, computational systems, and even the stability of the fundamental laws of physics.

4. Multiversal instability and dimensional paradoxes

Crossing between universes

If parallel universes exist, reckless interaction could alter events and reshape history in unpredictable ways.

Even minor interference – such as introducing foreign technology or biology – could destabilize ecosystems, political systems, and evolutionary paths.

The worst-case scenario would be a reality fracture, where fundamental constants shift randomly across dimensions. This would extinguish stars, and destroy planets and civilizations – and the damage could never be undone.

Temporal experiments and causal disasters

Time manipulation poses perhaps the most perilous ethical dilemma of all. Careless alterations could create self-erasing paradoxes or even rewrite history entirely.

Unregulated time experiments could fragment reality into unstable timelines, causing memory inconsistencies, temporal loops, or the collapse of coherent history across entire civilizations.

5. Preventing disaster on a universal scale

These scenarios highlight the need for extreme caution in experimental physics. Interstellar laws must regulate high-risk research and ensure that containment protocols are in place before experiments begin. Projects should operate within tightly controlled parameters, with risk assessments weighing potential benefits against even the slightest chance of existential catastrophe.

Some experiments might even be permanently prohibited because the risks associated with a single miscalculation are simply too high.

The ethical implications of manipulating fundamental physics

As we advance toward space-time engineering, quantum manipulation, and the harnessing of cosmic-scale energies, we will face some of the most profound ethical dilemmas in our history. The ability to modify gravitational forces, tap into dark energy, or alter universal constants will demand that we confront questions of responsibility, unintended consequences, and existential risk.

1. Cosmic responsibility and existential risks

Potentially irreversible changes to the universe

Mastering the deepest forces of physics offers astonishing opportunities, but it also carries immense danger. If we extract vacuum energy improperly or bend space-time beyond safe thresholds, the effects could extend far beyond a laboratory or test site. We might disrupt entire planetary systems, accelerate cosmic entropy, or destabilize the delicate equilibrium that governs the universe's expansion.

Even if these scenarios remain theoretical, they illustrate the need for extreme caution.

- What if a failed attempt to manipulate space-time causes a localized gravitational collapse that sends shockwaves across interstellar space?
- Could shifting quantum states on a cosmic scale fracture reality itself, unleashing chaos in the fundamental forces that shape existence?

To safeguard the fabric of the universe, we will need strict protocols that govern all experimentation with cosmic forces. No project should be allowed to proceed if it carries *any* credible risk of threatening *any* civilization – including those we have yet to discover.

The morality of engineering space-time

While controlling fundamental physics could give us unlimited energy, interstellar travel, and the ability to survive anywhere, the ethics of such power are complex. Do we have the right to reshape the universe for our own benefit if it risks harming other cosmic beings?

Our interference could disrupt alien civilizations – including those that have adapted to their own versions of physics in ways we cannot comprehend.

Modifying gravitational fields, harnessing dark energy, or altering space-time could destabilize planetary balances, interfere with survival mechanisms, and even alter evolutionary paths on worlds we have never seen.

Ethical safeguards must ensure that our power to reshape reality never comes at the expense of universal equilibrium or the integrity of existence.

2. Exploiting dark energy and universal resources

The ethics of energy monopolization

If we learn to extract dark energy or manipulate cosmic inflation, a critical question arises: should any single civilization be allowed to control such a force, or should it be shared universally?

Dark energy, for example, permeates the fabric of space-time. If we monopolized it, it could create extreme technological disparities, enabling one dominant species to wield power over all the others. This could fracture interstellar cooper-ation, trigger conflicts over access, and destabilize regions that depend on shared equilibrium.

To prevent this, we will need universal treaties to ensure that dark energy is used ethically, sustainably, and without disrupting natural cosmic processes.

Enforcing these agreements will require interstellar regulatory bodies where multiple civilizations collaborate to define limits and responsibilities.

But we must remember that humanity might not be the first to reach this threshold. Other species might already have this capability – and they might not have consulted us.

The consequences of controlling cosmic expansion

The ability to influence the universe's rate of expansion would be one of the most staggering powers imaginable. Dark energy drives the accelerating separation of galaxies, but what if we could slow it, halt it, or reverse it?

Slowing the expansion of the universe could stifle the formation of new star systems and alter galactic evolution. Accelerating it could fling entire regions into isolation, cutting off emerging civilizations from interstellar contact.

Decisions like these should not be left to a single species. Interstellar councils should bring together multiple civilizations to study, regulate, and ethically assess any attempt at cosmic engineering. They should also ensure that large-scale experiments cannot compromise the universe's integrity.

3. The ethics of multiversal manipulation

Crossing ethical boundaries into other universes

If the multiverse exists and we gain the ability to traverse it, we could face numerous unprecedented responsibilities. Alternate universes might have very different laws of physics, or branching timelines with unique histories.

Would it be ethical to alter events in another reality for our own benefit? Could our presence disrupt their natural evolution, erase civilizations, or derail their progress? Just as we follow diplomatic protocols when encountering new cultures on Earth, we will also need interdimensional protocols to prevent reckless interference.

If alternate civilizations exist, should we act only as observers? And if engagement becomes necessary, how do we uphold interdimensional sovereignty and respect their autonomy? It will be essential that we establish scientific and diplomatic guidelines before we cross into new realities and cause unintended harm.

The consequences of quantum consciousness transfer

Post-physical existence could offer us immense survival advantages, and free us from planetary dependence. But we must preserve autonomy and integrity of consciousness.

If the human mind becomes purely informational, we will need safeguards to protect individuality, emotional depth, and moral reasoning. Without them, we risk becoming nothing more than a vast computational network. And we will lose the essential qualities that make us human.

As we expand outward, we should also look inward, to ask what we leave behind, and what it means to be remembered.

Part VI

Legacy

A probe glides silently through the interstellar night.

A sculpture turns slowly in a ring's thin light.

A radio signal sweeps a thousand times across a quiet sky.

Legacy is less a monument than a message:
we were here, and we cared to say so.

15

Leaving a mark

We may send interstellar archives, monuments, and cultural artifacts into the cosmos – messages to whoever might find them. These could preserve our science, art, and history for trillions of years.

Leaving a legacy

In the vast scale of the universe, humanity might be but a momentary spark – an ephemeral presence in an infinite expanse. In the familiar thought experiment where the age of the universe is compressed into a single day, humanity appears only in the final second before midnight.

And yet, we have the urge to leave something behind; to carve our existence into the fabric of space and time.

> The thought experiment always ends at one second to midnight. But what would happen if the clock ticked on for one more second and actually reached midnight?
> Would it stop?
> Or would it go beyond and begin a new day?

Whether humanity endures for millions of years or fades into oblivion, our instinct to safeguard our knowledge, creativity, and influence is a declaration to those who might come after us:

- We existed.
- We questioned.
- We dreamed.
- We contributed.
- We mattered.

To preserve our thoughts, discoveries, and stories is to ensure that, even if we vanish, the echoes of our journey remain. Those echoes might take the form of interstellar archives, encoded transmissions, or artifacts drifting through deep space, waiting to be encountered by minds that never knew us.

1. Cosmic archives

Interstellar time capsules

One of the most ambitious ways to preserve human knowledge is through deep-space archives. We could launch probes carrying literature, art, philosophy, and scientific records beyond Earth. These floating libraries would carry the essence of our civilization to whoever might find them – whether they are extraterrestrial beings or our own far-flung descendants.

We have already taken small steps in this direction: the two *Voyager* spacecraft, launched in 1977, carry golden records etched with sounds, images, and messages from Earth. Future missions might carry vast data repositories containing the sum of our accumulated wisdom.

Some concepts are even more speculative. For example:

- Storing encoded knowledge in the gravitational orbits of black holes, using their immense stability as cosmic vaults.

- Preserving digital consciousness in quantum substrates. These living archives of human thought, creativity, and memory could share our stories and insights with future civilizations.

Galactic record-keeping

Unlike static records, galactic archives could be dynamic – continuously updated and transmitted across space.

AI-driven data stations could serve as broadcast hubs, sharing our scientific discoveries, cultural milestones, and historical events.

Off-world vaults on exoplanets or distant moons could safeguard our history, even if Earth ceases to be our home.

Data embedded in light waves, gravitational waves, or high-energy particles could radiate outward for millennia. They would turn the entire universe into a living archive of our achievements.

2. Artistic and cultural legacy beyond humanity

Humanity's legacy is more than science and technology; it is also the art, stories, and emotions we imprint on the cosmos.

Cosmic artifacts

Art preserves emotion, identity, and thought.

Imagine sculptures orbiting distant moons – silent monuments that tell the story of human civilization.

Imagine interstellar landmarks – anchored to asteroids, woven into planetary rings, or set adrift in interstellar space – carrying inscriptions, soundscapes, or visual archives that endure long after we are gone.

Just as ancient ruins, hieroglyphs and petroglyphs outlasted their creators by thousands of years, our monuments could speak for us across the ages.

Artificial intelligence could further extend our cultural immortality. It could continue to compose music, write great literature, and create art inspired by our traditions – long after our biological presence has faded.

Humanity's influence on the future of the cosmos

Our history is the story of a species capable of great wisdom and deep flaws. We scale unimaginable heights of knowledge while wrestling with the consequences of our actions. If we leave records for others to find, should they serve as guidance and inspiration, or as cautionary tales? Should we tell of our resilience, creativity, and scientific triumphs, or of the mistakes we made and the crises we failed to avert?

The beauty of transience

Human civilization might be brief in cosmic terms, but its influence will resonate far beyond its lifespan.

Like short-lived stars that burn brightly before fading, our impact might be fleeting yet profound.

Our discoveries in science, mathematics, philosophy, art, and literature have already altered the structure of knowledge and inspired generations. Even if no future civilizations ever find our archives, read our books, or hear our music, the universe has felt our presence. Our story was told. And that, in itself, is an extraordinary triumph.

Ensuring the continuity of human civilization

Alongside art, religion, and philosophy, other fundamental pillars of human civilization will endure. These will shape our identity, social structures, and governance as we advance technologically and expand into space. They will preserve the essence of our society across future generations and in environments far removed from those we know today.

Language and communication

Language remains our most powerful tool for transmitting knowledge, emotions, and cultural heritage. As we venture beyond Earth, communication will evolve. Out of necessity, it will adapt to the realities of life in deep space, while continuing to define how different civilizations express their identity, maintain connections, and share their stories.

Science and technological innovation

Human progress has always been driven by science and discovery. It reshapes how we understand the universe, extend our lifespans, and define intelligence. In an era of advanced technology, space exploration, artificial intelligence, and new energy systems, society will be transformed – perhaps to the point where we (yet again) need to redefine what it means to be human.

Governance and social organization

Space-based societies will need governance models that balance autonomy, cooperation, and ethical oversight. Stability will depend on laws, rights, and economic principles that can adapt to unprecedented challenges. As technology advances, so too will the frameworks that shape the structure of future civilizations.

Economics

In interstellar civilizations, traditional currency-based economies may become less relevant. Instead, resource-driven systems, optimized by AI, might take their place.

On Earth, financial wealth often dictates power. But in space, energy, materials, and technological capability might form the foundations of economic strength. Automated, AI-managed economies, powered by quantum computing, could allocate resources, optimize production, and accurately predict consumption patterns, ensuring both survival and efficiency.

Value might be linked to stored or transferable energy – such as fusion energy credits or reserves of antimatter. Access to critical resources – such as water, or rare metals and minerals from asteroids – could shape trade dynamics, influence governance, and serve as leverage in interstellar negotiations.

Interstellar trade networks could emerge, exchanging goods, raw materials, services, and knowledge across star systems. While these resources might not serve as currency in the traditional sense, they could underpin economic agreements and policy.

Social bonds and human relationships

No matter how advanced we become, certain aspects will remain central to our identity, including relationships, families, emotional fulfillment, and shared experiences.

These will continue to form the bedrock of our community, history, and culture, regardless of whether we remain biological, merge with AI, or transition into entirely new forms of existence.

Even among distant stars or within digital realities, our fundamental need for connection will endure, And this will shape the evolution of our societies beyond Earth.

Ethics and morality

As we expand further into space, the ethical dilemmas will grow more complex and far-reaching.

They will influence how we treat new worlds, interact with artificial intelligence, engage with cybernetic or digital life forms, and respond to the discovery of intelligent extraterrestrial beings.

New interstellar civilizations will need to be built on strong ethical foundations, ensuring that our progress does not come at the expense of responsibility.

But legacy is not always visible. Sometimes, it lies in the quiet traces. This is the influence that endures without recognition.

16

The unrecognized legacy

We often think of legacy as remembrance – a mark left on history through our achievements, ideas, and influence. But what if that legacy goes unrecognized? What if humanity – or any individual, for that matter – were to vanish without being acknowledged? Would their impact still have meaning? Does meaning depend on recognition, or does it reside in the act of creation?

Even if no conscious observer sees or remembers our discoveries, thoughts, or stories, our influence will, nevertheless, persist beyond the reach of our history books and monuments. The atoms we move, the energy we create, the ideas we express – all of these things ripple through reality, and subtly alter the course of events.

A forgotten scientist's work may never be cited, yet his equations might quietly inspire future breakthroughs. A piece of music might go unheard by the masses, but it might transform the mind of the one person who hears it, and it could influence artistic expressions yet to come.

Perhaps legacy holds meaning only because we assign value to it, not because the universe demands it. Or perhaps it is not about being remembered at all, but about having existed, having contributed, and having left something behind. Significance might not require remembrance; the act of trying, creating, and influencing can be meaningful in itself, regardless of whether it is recorded or remembered.

The universe does not need to acknowledge an action for it to have occurred. A person's life might be recognized and celebrated, or it might hardly be unnoticed. But it still contributed to the movement of existence – perhaps not in memory, but certainly in participation. From this perspective, legacy is not about permanence, but about *presence.*

Our imprint on the cosmos does not require validation; it simply exists. It exists in the molecules we rearranged through our creations, in the light waves that were altered by our machines, and in the interstellar transmissions we sent into the void. These things may persist for millennia, even if no conscious mind ever sees or deciphers them.

Ultimately, our story might never be read, but it will have been written. And in the grand unfolding of the universe, that, perhaps, is enough.

And so we arrive at our final question: if we can go anywhere and become anything, what story do we choose to tell?

Epilogue

The story we choose to tell

Resilience and adaptability: the strengths that define us

Our greatest strength has always been our ability to reinvent ourselves. Human civilization is not defined by its permanence, but by its capacity to evolve in response to changing conditions. Across history, human societies and empires have risen and fallen, yet we endure. We have transformed, certainly, but we still carry forward the ideas, innovations, and knowledge of those who came before us.

Shaping the future through transformation

Technological revolutions have repeatedly reshaped our lifestyles, economies, and social structures. The Industrial Revolution mechanized production and transformed labor and commerce. The Digital Revolution connected the world in an instant, creating new cultural landscapes and redefining how we share information.

Today, the rise of artificial intelligence and automation is once again changing how we work, create, and interact, pushing us into an era of unprecedented connectivity and capability. These massive changes – once seen as radical – are now essential to human progress. They demonstrate our remarkable ability to not only adapt to technological change, but to shape it.

The next great transformation

Our next giant leap forward may come from space exploration: colonizing Mars, building vast orbital habitats, and venturing toward distant exoplanets. Just as past civilizations adapted to Earth's changing landscapes and technologies, future generations will need to adjust to alien environments, self-sustaining ecosystems, and space-based economies.

Our greatest safeguard against existential threats will be scientific discovery. Advances in quantum physics, biotechnology, and interstellar engineering will give us the tools to adapt, build, and flourish in unfamiliar worlds. These tools will make deep-space habitation both possible and sustainable.

Beyond the biological

Digital intelligence could utterly transform consciousness. The ability to transfer our minds into AI systems or neural networks would enable us to transcend our physical limitations entirely. Future civilizations might exist only as digital entities, freed from illness, aging, and death.

If our intelligence, creativity, curiosity, and self-awareness remain intact, then our lack of physical form might not matter. Yet such a transformation would bring profound ethical dilemmas. Questions of identity, rights, and sentience could challenge our deepest understanding of what it means to be alive. And, inevitably, they would also force us to reconsider what it means to be human.

When life on Earth ends – whether in centuries, millennia or eons – our survival will depend on our willingness to adapt, to cooperate, and to imagine futures somewhere else. Along the way, we will face choices that define not only where we go, but who we become.

We might spread across the stars, transform ourselves beyond recognition, and leave behind traces that outlast the galaxies – or even the universe. Or we could simply vanish, our story told but unread.

But we have seen that meaning is not only found in remembrance; it is found in the act of striving. It is found in our courage, and in our drive to explore. It is in our will to endure, and in our choice to act with integrity, even when no one is watching.

The future will not be given to us; it will be *made* by us. And in making it, *we* will decide what it means to be human – not just on Earth, but wherever we choose to go.

Our drive to explore and survive

The course of human history has been shaped by discovery, adaptation, and perseverance. We have crossed vast oceans, set foot on the Moon, and now we are planning even greater voyages to the stars. We carry within us an inherent urge to push boundaries and uncover new possibilities.

1. Exploration – our need to go beyond

Our drive to explore is one of our most defining traits, and it is central to our identity as a species. It fuels our scientific breakthroughs and technological advances, as well as our bold journeys into the unknown.

The great migrations of early humans led to the settlement of diverse landscapes and the creation of interconnected civilizations. That same instinct to venture outward remains as strong today as it was thousands of years ago – except now, our gaze is fixed on interstellar space.

Curiosity and scientific discovery

Since our earliest days, our curiosity has driven us to observe the sky, map the stars, and seek to understand the universe. Ancient astronomers charted celestial movements, and laid the foundations for modern space science. Today, modern observatories and space telescopes allow us to peer across billions of light-years, revealing exoplanets, distant galaxies, and cosmic structures that were previously unknown.

Cutting-edge tools, including gravitational wave detectors and next-generation radio telescopes, are enabling us to study black holes, neutron stars, and the nature of space-time. As well as expanding our understanding, these advances open the door to possibilities such as interstellar travel and perhaps even the discovery of extraterrestrial life.

And, of course, our search for alien life and its origins remains one of our most profound scientific endeavors. How – and where – did life begin? Is Earth unique, or are there countless other worlds teeming with biology?

Upcoming missions to Mars, Europa, Titan and beyond aim to answer these questions, hunting for organic molecules and biosignatures. Meanwhile, initiatives like SETI and deep-space probes continue to listen for signals from other intelligences, reinforcing our desire to connect with the unknown.

2. Survival: adaptability in the face of challenges

Resilience through hardship

Throughout history, humanity has endured harsh climates, pandemics, economic collapses, and political upheavals. Each challenge has sparked ingenuity and transformation. In response to natural disasters, for example, we engineered novel flood defense systems, earthquake-resistant structures, and advanced forecasting systems to protect lives and rebuild communities.

Our adaptation to climate change has driven breakthroughs in sustainable agriculture, renewable energy, and atmospheric modification – proof that we can reshape civilization in the face of adversity. Science and technology remain our most powerful tools for countering existential threats. We now have the ability to eradicate diseases, extend our lifespans, and build protective infrastructures and self-sustaining habitats.

But our adaptability also extends well beyond Earth. Space habitats, biospheres, and off-world colonies could soon become the next frontier for human survival.

Preparing for long-term continuity

Establishing colonies on Mars, the Moon, and in deep-space habitats would diversify our survival strategy, and reduce our dependence on a single planet.

Multi-planetary existence would act as an insurance policy for our civilization, ensuring continuity no matter what challenges arise.

Our future survival will be shaped by both biological and technological evolution, including biomechanical augmentation, neurological adaptation, and AI-assisted resilience.

3. The future – exploration as survival

Moving beyond Earth for sustainability

Colonizing other planets and moons will give us multiple safety nets, ensuring that our civilization persists even if Earth becomes uninhabitable.

Once interstellar travel becomes feasible, we will begin to extend our reach beyond the Solar System, exploring habitable exoplanets and constructing interstellar habitats.

Generational ships, cryogenic stasis, and perhaps even vessels equipped with warp drives could one day allow our descendants to settle on multiple planetary systems. Together, they will create a network of civilizations that span the galaxy – and maybe far beyond it.

Ultimately, our survival depends on our adaptability – whether that is physical, intellectual or ethical. Exploration will no longer be solely about discovery; it will become the foundation of our enduring legacy.

Embracing the unknown

Exploration is more than a physical journey. It is a process of learning, transformation, and deepening our connection to the universe. It reshapes our thoughts, our identity, and our sense of responsibility. To explore is to act with courage. And we will need *immense* courage to confront uncertainty, embrace change, and challenge the assumptions we hold about our place in the cosmos.

The philosophical role of exploration

Every mission into deep space has the potential to uncover new principles that could transform our understanding of physics. It could also open pathways to explore further than we ever imagined. Space exploration is not merely an escape from Earth's limitations; it is a gateway to unlocking the deepest secrets of existence.

Each discovery has the power to reshape us in ways we cannot yet predict. History shows that our greatest breakthroughs came when we refused to fear the unknown and chose instead to pursue discovery – despite the risks.

The endless possibilities of the cosmos now await us, and they will inspire new generations of explorers, dreamers, and pioneers. And those generations will, in turn, define our amazing future among the stars.

Ellis

x

Bibliography and resources

Benford, J., Benford, G., & Benford, D. *Searching for Cost-Optimized Interstellar Beacons.* Astrobiology, 2010.

Bostrom, N. *Superintelligence: Paths, Dangers, Strategies.* Oxford University Press, 2014.

Ćirković, M. M. *The Astrobiological Landscape: Philosophical Foundations of the Study of Cosmic Life.* Cambridge University Press, 2012.

Dyson, F. J. *Search for Artificial Stellar Sources of Infrared Radiation.* Science, 1960.

Harrison, A. *After Contact: The Human Response to Extraterrestrial Life.* Routledge, 1991.

Hawking, S. *Brief Answers to the Big Questions.* Bantam, 2018.

Hein, A. M. et al. *Project Dragonfly: Design of an Interstellar Mission.* Acta Astronautica, 2014.

Hennessey, J. L. (Ed.). *History and Speculative Fiction.* Springer, 2024.

Johnson, C. T. *Humanity's Fate: An Analysis of Speculative Human Evolution in Literary Fiction.* University of South Carolina, 2023.

Kaku, M. *The Future of Humanity: Terraforming Mars, Interstellar Travel, Immortality, and Our Destiny Beyond Earth.* Doubleday, 2018.

Kardashev, N. S. *Transmission of Information by Extraterrestrial Civilizations.* Soviet Astronomy, 1964.

NASA. *Space Colonization Bibliography.* NASA Headquarters Library.

Neukart, F. *Toward the Stars: Technological, Ethical, and Sociopolitical Dimensions of Interstellar Exploration.* arXiv preprint, 2024.

Rees, M. *Our Final Century.* Penguin, 2003.

Romanovskaya, I. K. *Migrating Extraterrestrial Civilizations and Interstellar Colonization: Implications for SETI and SETA.* International Journal of Astrobiology, Cambridge University Press, 2022.

Sagan, C. *Cosmos.* Random House, 1980.

Smith, K. *The Ethics of Space Exploration.* Springer, 2016.

Stapledon, O. *Star Maker.* Methuen, 1937.

Tipler, F. J. *Extraterrestrial Intelligent Beings Do Not Exist.* Quarterly Journal of the Royal Astronomical Society, 1980.

Trocha, B. *Speculative Fiction for the Future of Man and Civilization.* Future Human Image Journal, 2019.

Turchin, A. *Risks of Human Extinction in the 21st Century.* Futures, 2018.

Vint, S. *Biopolitical Futures in Twenty-First-Century Speculative Fiction.* Cambridge University Press, 2021.

Willems, B. *Speculative Realism and Science Fiction.* Edinburgh Scholarship Online, 2017.

Zubrin, R. *The Case for Mars.* Free Press, 1996.

The Future of Interstellar Exploration. Simple Science, 2025.

About the author

Dr. Ellis Silver is an American-born ecologist and environmentalist who specializes in aquatic habitats.

He currently works as a developmental engineer and researcher on ocean-cleanup vessels in Southeast Asia.

He is happy to answer sensible questions by email – on the rare occasions when he has a decent internet connection.

Contact address: silver.ellis@gmail.com

You might also like to follow his Facebook page www.facebook.com/HumansAreNotFromEarth where he and his team post interesting news and articles related to his books and research.

Acknowledgments

The author is once again hugely grateful to Dave Haslett and the team at ideas4writers for their work in turning his hundreds of pages of rough notes into this book.

He would also like to thank the tens of thousands of people who have read his books, and the many hundreds who have taken the time to send him messages.
He apologizes for not being able to respond to every message.

***

The editorial team acknowledges the use of AI in adapting Dr. Silver's notes into this book. It reduced a near-impossible three-year task to around six months.

Index

www.ingramcontent.com/pod-product-compliance
Lightning Source LLC
LaVergne TN
LVHW052217200926
848481LV00001B/131